T0231180

Fractal Patterns in Nonlinear Dynamics and Applications

Santo Banerjee
Institute for Mathematical Research
University Putra Malaysia
Serdang, Malaysia

M K Hassan
Dhaka University
Dhaka, Bangladesh

Sayan Mukherjee
Department of Mathematics
Sivanath Sastri College
Kolkata, India

A Gowrisankar
Department of Mathematics
Vellore Institute of Technology
Vellore, Tamil Nadu, India

CRC Press
Taylor & Francis Group
Boca Raton London New York

CRC Press is an imprint of the
Taylor & Francis Group, an **informa** business

A SCIENCE PUBLISHERS BOOK

CRC Press
Taylor & Francis Group
6000 Broken Sound Parkway NW, Suite 300
Boca Raton, FL 33487-2742

First issued in paperback 2021

© 2020 by Taylor & Francis Group, LLC
CRC Press is an imprint of Taylor & Francis Group, an Informa business

No claim to original U.S. Government works

Version Date: 20191014

ISBN-13: 978-1-03-208351-3 (pbk)
ISBN-13: 978-1-4987-4135-4 (hbk)

**Visit the Taylor & Francis Web site at
http://www.taylorandfrancis.com**

**and the CRC Press Web site at
http://www.crcpress.com**

*This volume is dedicated to all researchers
who love to explore the mystery of nature*

Preface

Mathematics and Physical Sciences have been used together to describe natural phenomena. As a result of its numerous successes, there have been a growth of scientific discoveries. This book is meant for anyone who wants to understand patterns of fractal geometry in detail along with the physical aspects and basic mathematical background. It is our goal to give readers a broad interpretation of the underlying notions behind fractals and multifractals. Furthermore, we want to illustrate the fundamentals of fractals, stochastic fractals and multifractals with applications.

Many phenomena in nature exhibit self-similarity. That is, either a part is similar to the whole or snapshots of the same system at different times are similar to one another albeit it differs in size. Initially this book describes novel physical applications and the recent progress through scale-invariance and self-similarity. In general, mathematics is concerned with sets and functions to model real world problems which are done by classical Euclidean geometry. However, there are many phenomena which are traditionally observed as too irregular or complex to be described using classical Euclidean geometry. In such a case, there is a need for alternative geometry to resolve these complexities which helps in a better understanding of natural patterns. The idea of irregular objects has been revolutionized by Benoit B Mandelbrot and is called fractal geometry. It has generated a widespread interest in almost every branch of science. The advent of inexpensive computer power and graphics has led to the study of non-traditional geometric objects in many fields of science and the idea of the fractal has been used to describe them. In a sense, the idea of fractals has brought many seemingly unrelated subjects under one umbrella. The second chapter deals the construction of fractals through an iterated function system of contractive mappings and illustrates some examples.

Nature loves randomness and natural objects which we see around us evolve in time. The apparently complex look of most of natural objects does not mean that nature favours complexity, rather that the opposite is true. Often the inherent and the basic rule is trivially simple; it is in fact the randomness and the repetition of the same simple rule over and over again that makes the object look complex. Of course, natural fractals cannot be strictly self-similar rather they are statistically self-similar. For instance, one can draw a curve describing the tip of the trees in the horizon but the details of the two pictures drawn by the same person will never be the same despite how hard one tries. Capturing the generic feature of cloud distribution without knowing anything about self-similarity can be described as our natural instinct. The subsequent section attempts to incorporate both the ingredients (randomness and kinetic) to the various classical fractals, namely stochastic fractals and multifractals, in order to know what role these two quantities play in the resulting processes.

Contents

Symbols

$\mathbb{N}$	the set of all natural numbers.
$\mathbb{R}$	the set of all real numbers.
$\mathbb{R}^n$	the n-dimensional Euclidean space.
(X, d)	metric space.
$\mathcal{K}(X)$	collection of all non-empty compact subsets of X.
H_d	Hausdorff metric induced by the metric d.
$\mathcal{N}(K, \varepsilon)$	smallest number of closed balls required to cover K.
dim_T	topological dimension.
dim_B	box dimension or fractal dimension.
$\mathcal{H}^s(K)$	s-dimensional Hausdorff measure of K.
dim_H	Hausdorff dimension.
RE_q	Renyi entropy of order q.
D_q	generalized fractal dimension of order q.
a_0	the Bohr radius.
$P(m, N)$	probability that the walker is at position m after N random steps.
$\mathcal{C}[a, b]$	set of all continuous real-valued functions defined on the closed interval $[a, b]$.
$B(x, r)$	open ball centred at x with radius r.
$B[x, r]$	open ball centred at x with radius r.
$\tau(q)$	the mass exponent.
$M(m, n; t)$	2-tuple Mellin transform.
D_0	the dimension of the support or fractal dimension.
D_1	the information dimension.
D_2	the correlation dimension.

Chapter 1

Scaling, Scale-invariance and Self-similarity

Physics is all about observations and the measurement of physical quantities with the desire to find patterns in the data. These patterns subsequently lead to finding principles of the phenomena under investigation. Finding patterns often means finding order, similarity, self-similarity and scale-invariance in the phenomena under investigation which may otherwise look disordered. Many phenomena in nature exhibit similarity and self-similarity. That is, either part is similar to the whole or snapshots of the same system at different times are similar to one another albeit they differ in sizes. Trying to understand the idea of self-similarity is therefore like trying to learn their nature. It has proved quintessential to understand the idea of dimension and dimensionless quantity to really grasp the idea of similarity and self-similarity.

The building blocks of physics are the physical quantities quantified by numbers which are obtained by measurements. When we can measure a physical quantity and express it in numbers only then we can say we have gained some knowledge about that quantity; but when we cannot express it in numbers then our knowledge about that quantity can be described as meagre and unsatisfactory. Numbers to quantify physical quantities are typically obtained by a direct or indirect comparison with the corresponding units of measurements which are of two kinds and they are called fundamental and derivative units. The fundamental units of measurements are usually defined arbitrarily in the form of certain standards, artificial or natural. The derivative ones, on the other hand, are obtained from the fundamental units of measurements by virtue of the definitions which always indicates some conceptual method or definition. For instance, speed is *the ratio of the distance traversed in a given time interval to the magnitude of that time interval* and hence as a unit of speed one can take the ratio of the unit of length to the unit of time in the given system. A set of fundamental units of measurement is sufficient for measuring the properties of a class of phenomena under investigation and it is then called a system of units of measurement. Until recently, the centimeter-gram-second system, abbreviated as CGS system, has been widely used by scientists working in laboratories, dealing with small quantities where one gram (gm) is adopted as the unit of mass, one centimeter (cm) is adopted as the unit of length, and one second (s) is adopted as the unit of time. A system of units of measurement consisting of two units such as a unit for the measurement of length and a unit for the measurement of time for example is sufficient for investigating the kinetic phenomena, while a system of units consisting of only one length unit is sufficient for measuring the geometric aspects of the object. In addition to the CGS system, one often considers a system in which 1 km (10^5 cm) is used as the unit of length, 1 metric ton (10^6 gm) is used as the unit of mass, and 1 hour (3600 s) is used as the unit of time. These two systems of units have the following property in common. The standard quantities and the fundamental units are of the same physical nature (mass, length, and time). Consequently, we say that these systems belong to the same class. To generalize, a set of systems of units that differ only in the magnitude (but not in the physical nature of their standard) of the fundamental units is called a class of systems of units.

The system just mentioned and the CGS system are members of the same class since the same standards of lengths, masses and time are used as the fundamental units. The corresponding units for an arbitrary system within the same class are as follows: (i) unit of length = cm/L, (ii) unit of mass = g/M, and (iii) unit of time = s/T, where L, M, T are the abstract positive numbers that indicate the factors by which the fundamental units of length, mass and time decreases in passing from the original system (in this case, the cgs system) to another system in the same class. This class is called the LMT class. In particular, we find that one of the widely used the *SI* systems belongs to the same *LMT* class, in which the unit of mass is taken to be 1 kg=1000 gm, which is complete mass of the previously mentioned standard of mass; the unit of length is taken to be 1 meter = 100 cm, which is the complete length of the standard of length mentioned before; and the unit of time is taken to be 1 second. Thus upon passage from the cgs system to the SI system $M = 0.001$, $L = 0.01$, and $T = 1$. Often one also uses systems of the FLT class, in which the fundamental units of measurement have the form kgf/F, cm/L, and s/T. These are certainly not the only possible choice of units of measurements rather the choice of units of measurements depends on the scale of the physical quantity under investigation. For instance in the study studying phenomena at the atomic and molecular level it is useful to choose a unit of length which is comparable to the size of an atom. We do this by using the electron charge e as the unit of charge, and the electron mass me as the unit of mass. At the atomic level the forces are all electromagnetic, and therefore the energies are always proportional to $\frac{e^2}{4\pi\epsilon_0}$ which has dimensions ML^3T^{-2}. Another quantity that appears in quantum physics is $\hbar$, the Plank's constant divided by 2π which has dimensions ML^2T^{-1}. Dimensional analysis then reveals that the atomic unit of length is

$$a_0 = \frac{\hbar^2}{\frac{m_e e^2}{4\pi\epsilon_0}}. \tag{1.1}$$

This is known as the Bohr radius or radius of smallest orbit for an electron circling around proton of the hydrogen atom. In the atomic world the Bohr radius can be chosen as the standard of length. In astronomy we deal with very large distances hence the Bohr radius as a standard can be highly inconvenient and therefore several other units are in use. For instance, the astronomical unit (AU = 1.496×10^{11} m) in which the average distance between the Earth and the Sun is taken as the standard. The basic idea is that the physical laws do not depend upon arbitrariness in the choice of the basic units of measurement. Newton's second law, $F = ma$, is true regardless of whether we measure mass in kilograms, acceleration in meters per second squared, and force in newtons, or whether we measure mass in slugs, acceleration in feet per second squared, and force in pounds. This universality of the physical laws makes science so beautiful.

The measurement systems are fundamental to physics as it allows us to obtain the numerical number for quantifying different physical quantities. However, the precise magnitude of the numerical value strictly depends on the choice of the class of units of measurements we use to obtain it. Any investigation in physics ultimately comes down to determining certain quantity which may depend on various other physical quantities that characterizes the phenomena under consideration. The problem therefore reduces to establishing the relationship which can always be represented in the form

$$f = f(a_1, ..., a_n), \tag{1.2}$$

where f is the quantity of primary interest which we will call the governed parameter and the quantities $a_1, a_2, ..., a_n$ on which f depends will be called governing parameters. The governing parameters are usually combined into one single term so that it bears the same physical dimension as the one for f. In this opening chapter we shall go over some important aperitifs that we will need throughout this book. We will discuss the idea of dimension of physical quantity and how it differs from dimensionless quantity. We will then rigorously prove that the dimensional function of the physical quantities are always power-law monomial in nature. The primary aim of this chapter is to establish the fact that the scale-invariance, self-similarity or scaling as well as the mathematical definitions of the homogeneous function are all deeply rooted to the power-law monomial nature of the dimensions of the physical quantity (recommended reading [10, 20, 32, 37, 39]).

1.1   Dimensions of physical quantity

It can be said with certainty that each and everyone reading this book have dealt with dimensional analysis at some stage of their studies, especially while solving problems in their early undergraduate years. Yet, many of us may not know how to separate a dimensionless quantity from a dimensional quantity? It is absolutely essential to know a clear-cut definition of the dimensional quantity and how it differs from dimensionless quantity. Dimension of a physical quantity is the function that determines the number of times one needs to alter the numerical value of this quantity while passing from one system of units of measurement to another system within the same class. For instance, if the unit of length is decreased by a factor L, the unit of mass is decreased by a factor of M and the unit of time is decreased by a factor T then the unit of force is smaller by a factor of MLT^{-2} than the original unit. Consequently, the numerical values of all the forces in the new unit of measurement will be increased by a factor of MLT^{-2} owing to the definition of the principle of equivalence. Upon decreasing the unit of mass by a factor M and the unit of length by a factor of L, we find that the new unit of density is smaller by factor ML^{-3} than the original unit, so that the numerical values of all the densities are increased by a factor ML^{-3}. One can treat the other physical quantities of interest in a similar fashion. Let us consider that we have a stick whose dimension function is $\phi(M, K, T) = L$. Say that we want to measure its length using a meter scale and found that its length is 9 meters. Say, now that we decreased the unit of measurement by a factor of $L = 100$ and hence in the new scale the numerical value will be increased by a factor of 100 since its dimension is L.

On the other hand, quantities whose numerical values remain identical in all systems of units within a given class are called *dimensionless* and hence the dimension function is equal to unity for a dimensionless quantity. Perhaps a couple of examples at this stage will provide a better understanding about the dimensionless quantity than its mere definition. Consider, that you want to measure the ratio of the circumference to the diameter of a circle. Regardless, of the choice of the unit of measurement the numerical value of the ratio will always be the same. We will find this definition extremely useful not only in this chapter but in the subsequent chapters as well. Again, consider that the angular frequency ω of a pendulum of length l for small oscillations is 9.8 cycle/sec. On the other hand using a simple dimensional analysis or detailed solution reveals that the natural frequency of a pendulum is $\sqrt{g/l}$ where g is the acceleration due to gravity, which is $9:8$ ms^{-2} on earth (in the SI system of units). To obtain the natural frequency of the simple pendulum one usually needs to solve the differential equation which results from applying Newton's second law to the pendulum. Thus it is clear that the ratio

$$\Pi = \frac{\omega}{\sqrt{g/l}}, \tag{1.3}$$

is a dimensionless quantity. Suppose that $l = 1$m and $\omega = 9.8$cycles/sec and hence $\Pi = \sqrt{9.8}$ in the SI system of units. Now let us change the system of units so that the unit of mass is decreased by a factor of $M = 1000$, the unit of length is decreased by a factor of $L = 100$, and the unit of time is decreased by a factor of $T = 1$. With this change, the units of frequency will decrease by a factor of $T^{-1} = 1$ and the units of acceleration by a factor of $LT^{-2} = 100$. Therefore, the numerical value of l in the system of units of measurement will be $l100$ instead of 1 and that of g will be 980. However, the numerical value of Π will still remain invariant under a change of units and hence it is a dimensionless quantity.

1.2 Buckingham Pi-theorem

The name Pi comes from the mathematical notation Π as defined here, for historical reason, to describe the dimensionless variables obtained from the power products of governing parameters denoted by Π_1, Π_2, Π_3..., etc. If a physical process involves investigation of a certain dimensional physical quantity (governed quantity) that depends on n number of other dimensional variables then the Buckingham Pi theorem guides us in a systematic way to reduce a function of n variable problems into a function of k dimensionless variable problems if each of the k dimensional variables of the original n variables can be expressed in terms of the $n - k$ dimensionally independent variables. Now the question is: What is dimensionally independent variable? A set of dimensional variables, $a_1, a_2, \ldots, a_k$ of the total n variables, is said to have independent dimensions if none of the k variables have dimensions which can be represented as a product of powers of the dimensions of the remaining variables. For instance, density $[\rho] = ML^{-3}$, velocity $[v] = LT^{-1}$ and force $[F] = MLT^{-2}$ have independent dimensions in the sense that none of these quantities can be expressed in terms of the rest. In other words, a product of powers of these quantities doesn't exist which is dimensionless. On the other hand consider the case where force is replaced by pressure $[P] = ML^{-1}T^{-2}$. In this case the dimension of pressure can be expressed in terms of the dimension of that of velocity and density since $[P] = [v^2][\rho]$ and hence velocity and density have independent dimension but no pressure.

The first step in modeling any physical phenomena is the identification of the relevant variables, and then relating these variables via known physical laws. For sufficiently simple phenomena, we can usually construct a quantitative relationship amongst the physical variables from the first principles; however, for many complex phenomena such an *ab initio* theory is often difficult, if not impossible. In these situations constructing a model in a systematic manner with minimum input parameters can be a quite useful method for dimensional analysis. In fact, it has been proved to be a quite powerful method for analyzing experimental data. Most undergraduates have already encountered dimensional analysis during their course studies but without knowing its full potential. Here we will use dimensional analysis to capture techniques for analyzing experimental or empirical data or to solve partial differential equations which also help infer some insights about the solution. As we have already stated the relationship found in physical theories or experiments can always be represented in the form

$$a = f(a_1, a_2, \ldots, a_n), \tag{1.4}$$

where the quantities $a_1, a_2, \ldots, a_n$ are called the governing parameters.

Any investigation ultimately comes down to determining one or several dependencies of the form Eq. (1.4). It is always possible to classify the governing parameters $a_1, ..., a_n$ into two groups using the definition of dependent and independent variables. Let the arguments $a_{k+1}, ..., a_n$ have independent dimensions and the dimensions of the arguments $a_1, a_2, ..., a_k$ can be expressed in terms of the dimensions of the governing independent parameters $a_{k+1}, ..., a_n$ in the following way:

$$[a_1] = [a_{k+1}]^{\alpha_1} \cdots [a_n]^{\gamma_1}, \tag{1.5}$$
$$[a_2] = [a_{k+1}]^{\alpha_2} \cdots [a_n]^{\gamma_2},$$

$$\cdots \quad \cdots \quad \cdots$$
$$\cdots \quad \cdots \quad \cdots$$

$$[a_k] = [a_{k+1}]^{\alpha_k} \cdots [a_n]^{\gamma_k}.$$

The dimension of the governed parameter a must also be expressible in terms of the dimensionally independent governing parameters $a_1, ..., a_k$ since a does not have an independent dimension and hence we can write

$$[a] = [a_{k+1}]^{\alpha} \cdots [a_n]^{\gamma}. \tag{1.6}$$

Thus, there exists numbers α, β, and γ, such that Eq. (1.6) holds. We can therefore have a set of dimensionless governing parameters

$$\Pi_1 = \frac{a_1}{[a_{k+1}]^{\alpha_1} ... [a_n]^{\gamma_1}}, \tag{1.7}$$

$$\Pi_2 = \frac{a_2}{a_{k+1}^{\alpha_2} ... a_n^{\gamma_2}},$$

$$\cdots \quad \cdots \quad \cdots$$

$$\Pi_k = \frac{a_k}{[a_{k+1}]^{\alpha_k} ... [a_n]^{\gamma_k}},$$

and a dimensionless governed parameter

$$\Pi = \frac{f(\Pi_1([a_{k+1}]^{\alpha_1} ... [a_n]^{\gamma_1}), ..., \Pi_k([a_{k+1}]^{\alpha_k} ... [a_n]^{\gamma_k}, a_{k+1}, ..., a_n)}{[a_{k+1}]^{\alpha} ... [a_n]^{\gamma}}. \tag{1.8}$$

The right hand side of this equation clearly reveals that the dimensionless quantity Π is a function of $a_{k+1}, ..., a_n, \Pi_1, ..., \Pi_k$, i.e.,

$$\Pi \equiv F(a_{k+1}, ..., a_n, \Pi_1, ..., \Pi_k). \tag{1.9}$$

The quantities $\Pi, \Pi_1, ..., \Pi_k$ are obviously dimensionless, and hence upon transition from one system of units to another inside the given class their numerical values must remain unchanged. At the same time, according to the above, one can pass to a system of units of measurement such that any of the parameters of $a_{k+1}, ..., a_n$, say for example, a_{k+1}, is changed by an arbitrary factor, and the remaining ones are unchanged. Upon such a transition the first argument of F is changed arbitrarily, and all the other arguments of the function remain unchanged as well as its value Π. Hence, it follows $\frac{\delta F}{\delta a_{k+1}} = 0$ and entirely analogously $\frac{\delta F}{\delta a_{k+2}} = 0, \dots \frac{\delta F}{\delta a_n} = 0$. Therefore, the relation Eq. (1.9) is in fact represented by a function of k arguments and proves it is independent of $a_{k+1}, \dots, a_n$, i.e.,

$$\Pi = \Phi(\Pi_1, \dots, \Pi_k), \tag{1.10}$$

and the function f can be written in the following special form

$$f(a_1, \dots, a_k, \dots, a_n) = a_{k+1}^\alpha \cdots a_n^\gamma \Phi((\Pi_1, \dots, \Pi_k). \tag{1.11}$$

This constitutes the content of the central statement of dimensional analysis that has got far reaching consequences in understanding the scaling theory. It is also known as the π-theorem that has been formulated and proved for the first time by E. Buckingham.

The importance and use of dimensional analysis and the Buckingham π-theorem in particular in the description of the scaling or self-similarity can hardly be exaggerated. Theoretically, it is found that most physical systems are governed either by partial differential or by partial integro-differential equations and finding scaling or self-similarity will be the subject of intense and detailed consideration in the subsequent chapters. Nevertheless, here we give some instructive examples that will certainly help with grasping the essential ideas behind the π-theorem. In particular, the definition of dimensionless quantity will play the central role in establishing scaling or self-similarity.

1.3 Examples to illustrate the significance of Π-theorem

Perhaps the simplest and yet the most amusing example to demonstrate how the idea of the π-theorem can be used to prove the Pythagorean theorem. Consider a right angle triangle where three sides are of size a, b and c and, for definiteness, the smaller of its acute angles θ. Assume that we are to measure the area S of the triangle. The area S can be written in the following form

$$S = S(a, b, c). \tag{1.12}$$

However, the dimension of two governing parameters a and b can be expressed in terms of c alone since we have

$$[a] \sim [c] \quad \text{and} \quad [b] \sim [c], \tag{1.13}$$

and so it is true for the governed parameter S as we can write the dimensional relation $[S] \sim [c^2]$. We therefore can define two dimensionless governing parameters

$$\Pi_1 = \sin\theta = a/c \quad \text{and} \quad \Pi_2 = \cos\theta = b/c, \tag{1.14}$$

and the dimensionless governed parameter

$$\Pi = \frac{S}{c^2} = c^{-2} S(c\Pi_1, c\Pi_2, c) \equiv F(c, \Pi_1, \Pi_2). \tag{1.15}$$

Now it is possible to pass from one unit of measurement to another system of unit of measurement within the same class and upon such transition the arguments Π_1 and Π_2 of the function F as well as the function itself remains unchanged. It implies that the function F is independent of c and hence we can write

$$\Pi = \phi(\Pi_1, \Pi_2). \tag{1.16}$$

However, Π_1 and Π_2 both depend on the dimensionless quantity θ, i.e.,

$$\phi(\Pi_1, \Pi_2) = \phi(\theta). \tag{1.17}$$

and hence we can write

$$S = c^2 \phi(\theta), \tag{1.18}$$

where the scaling function $\phi(\theta)$ is universal in character.

In order to further capture the significance of Eq. (1.18) we re-write it as $\frac{S}{c^2} \sim \phi(\theta)$. This result has far reaching consequences. For instance, consider that we have a right triangle of with arbitrary sides $a' \neq a$, $b' \neq b$ and $c' \neq c$ but have the same acute angle θ as before. This can be ensured by choosing an arbitrary point on the hypotenuse of the previous triangle and drop a perpendicular on the base b. Consider that the area of the new triangle is S', yet we will have

$$\frac{S}{c^2} = \frac{S'}{(c')^2}, \tag{1.19}$$

since the numerical value of the ratio of the area over the square of the hypotenuse depends only on the angle θ. It implies that if we plot the ratio of the area over the square of the hypotenuse as a function θ all the data points should collapse onto a single curve regardless of the size of the hypotenuse and the respective areas of the right triangle. In fact, the detailed calculation reveals that

$$\phi(b/c) = \frac{1}{4}\sin 2\theta. \tag{1.20}$$

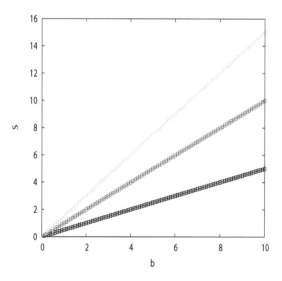

Figure 1.1: This figure shows how area S for one of the triangle of the above figure (say the largest triangle) varies as the base b is changed.

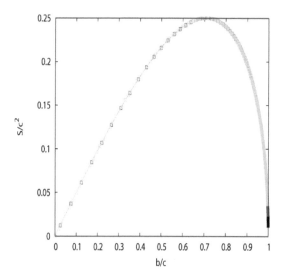

Figure 1.2: This shows that if we plot S/c^2 vs b/c instead of S vs b then all the three curves of Fig. (3) collapse on to a single master curve. It implies that for a given numerical value of the ratio b/c regardless of the size of the triangle corresponding S/c^2 is the same because triangle that satisfies these conditions are similar and hence the function $\phi(b/c)$ is unique.

Now the question is: what does this data collapse imply? It implies that if two or more right triangles which have one of the acute angles identical then such triangles are similar. We shall see later that whenever we will find a data collapse between two different systems of the same phenomenon then it would mean that the corresponding systems or their underlying mechanisms are similar. Similarly, if we find that data collected from the whole system collapsed with similarly collected data from a suitably chosen part of the whole system then we can conclude that the part is similar to the whole. On the other hand, if we have a set of data collected at many different times for a kinetic system and find they all collapse onto a single curve then we can say that the same system at different times are similar. However, similarity in this case is found of the same system at different times and hence we may coin it as temporal self-similarity.

Proof of the Pythagoras theorem: The area S of the original right triangle is determined by the size of its hypotenuse c and the acute angle θ since $S = c^2 \phi(\theta)$. Now, by drawing the altitude which is perpendicular to the hypotenuse c we can divide the original triangle into two smaller right triangles whose hypotenuses are a and b and say their respective areas are S_1 and S_2. The two smaller right triangles will also have their acute angle *theta* and hence we can easily show that

$$S_1 = a^2 \phi(\theta) \quad \text{and} \quad S_2 = b^2 \phi(\theta). \tag{1.21}$$

On the other hand we have

$$S = S_1 + S_2, \tag{1.22}$$

and it immediately leads to the conclusion that $c^2 = a^2 + b^2$ which is the Pythagorean theorem.

1.4 Similarity

In most cases, before a large and expensive structure or object such as a ship or aeroplane is built, efforts are always devoted to testing on a model system instead of working directly with the real system. The results of testing on model systems are then used to infer the various physical characteristics of the real object or structure under their future working conditions. In order to have successful modeling, it is necessary that we know how to relate the results of testing on models to the actual manufactured product. Needless to mention that if such connections are not known or cannot be made then modeling is simply a useless pursuit. For the purpose of rational modeling, the concept and the importance of the similarity phenomena can hardly be over emphasized. Indeed, the physically similar phenomena is the key to successful modeling.

The concept of physical similarity is a natural generalization of the concept of similarity in geometry. For instance, two triangles are similar if they differ only in the numerical values of the dimensional parameters, i.e., the lengths of the sides, while the dimensionless parameters, the angles at the vertices are identical for the two triangles. Analogously, physical phenomena are called similar if they differ only in their numerical values of the dimensional governing parameters; the values of the corresponding dimensionless parameters $\Pi_1, ..., \Pi_m$ being identical. In connection with this definition of similar phenomena, the dimensionless quantities are called *similarity parameters*.

Let us consider two similar phenomena, one of which will be called the prototype and the other the model; it should be understood that this terminology is just a convention. For both phenomena there is some relation of the form

$$a = f(a_1, a_2, a_3, b_1, b_2), \tag{1.23}$$

where the function f is the same in both cases by the definition of similar phenomena, but the numerical values of the governing parameters a_1, a_2, a_3, b_1, b_2 are different. Thus for prototype we have

$$a_p = f(a_1^{(p)}, a_2^{(p)}, a_3^{(p)}, b_1^{(p)}, b_2^{(p)}), \tag{1.24}$$

and for the model system we have

$$a_m = f(a_1^{(m)}, a_2^{(m)}, a_3^{(m)}, b_1^{(m)}, b_2^{(m)}), \tag{1.25}$$

where the index p denotes quantities related to the prototype and the index m denotes quantities related to the model. Consider that b_1 and b_2 are dependent variables so that they can be expressed in terms of a_1, a_2 and a_3 in both model and prototype systems. Using dimensional analysis we find for both phenomena

$$\Pi^{(p)} = \Phi(\Pi_1^{(p)}, \Pi_2^{(p)}), \tag{1.26}$$

and

$$\Pi^{(m)} = \Phi(\Pi_1^{(m)}, \Pi_2^{(m)}), \tag{1.27}$$

where the function Φ must be the same for the model and the prototype. By the definition of similar phenomena the dimensionless quantities must be identical in both the cases such as in the proto-type and in the model, i.e.,

$$\Pi_1^{(m)} = \Pi_1^{(p)}, \qquad \Pi_2^{(m)} = \Pi_2^{(p)}. \tag{1.28}$$

It also follows that the governed dimensionless parameter satisfies

$$\Pi^{(m)} = \Pi^{(p)}.$$ (1.29)

Returning to dimensional variables, we get from the above equation

$$a_p = a_m \left(\frac{a_1^{(p)}}{a_1^{(m)}}\right)^p \left(\frac{a_2^{(p)}}{a_2^{(m)}}\right)^q \left(\frac{a_3^{(p)}}{a_3^{(m)}}\right)^r,$$ (1.30)

which is a simple rule for recalculating the results of measurements on the similar model for the prototype, for which direct measurements may be difficult to carry out for one reason or another.

The conditions for similarity of the model to the prototype-equality of the similarity parameters Π_1, Π_2 for both phenomena show that it is necessary to choose the governing parameters $b_1^{(m)}, b_2^{(m)}$, of the model so as to guarantee the similarity of the model to the prototype:

$$b_1^{(m)} = b_1^{(p)} \left(\frac{a_1^{(m)}}{a_1^{(p)}}\right)^{\alpha_1} \left(\frac{a_2^{(m)}}{a_2^{(p)}}\right)^{\beta_1} \left(\frac{a_3^{(m)}}{a_3^{(p)}}\right)^{\gamma_1},$$ (1.31)

and

$$b_2^{(m)} = b_1^{(p)} \left(\frac{a_1^{(m)}}{a_1^{(p)}}\right)^{\alpha_2} \left(\frac{a_2^{(m)}}{a_2^{(p)}}\right)^{\beta_2} \left(\frac{a_3^{(m)}}{a_3^{(p)}}\right)^{\gamma_2},$$ (1.32)

whereas the model parameters $a_1^{(m)}, a_2^{(m)}, a_3^{(m)}$ can be chosen arbitrarily. The simple definitions and statements presented above describe the entire content of the theory of similarity: we emphasize that there is nothing more to this. We will now give a few examples to help us grasp the idea.

1.5   Self-similarity

The notion of self-similarity will be an underlying theme for the rest of the book. In a way the word self-similarity needs no explanation. Perhaps the best way to perceive the meaning of self-similarity is to consider an example and at this stage there is no better example than a cauliflower. The cauliflower head contains branches or parts, which when removed and compared with the whole is found to be very much the same except it is scaled down. These isolated branches can again be decomposed into smaller parts, which again look very similar to the whole as well as of the branches. Such self-similarity can easily be carried through for about three to four stages. After that the structures are too small to go for further dissection. Of course, from the mathematical point of view the property of self-similarity may be continued through an infinite stages though in real world such properties sustain only a few stages. Self-similarity can also be found on smaller scales. For instance, snowflakes often display self-similar branching patterns and aggregating colloidal particles have growth patterns that are statistically self-similar. Self-similarity, in fact, is so widespread in nature that they are exceedingly easy to find. Consequently, the main ideas of self-similarity are not difficult to grasp. Hopefully, when we are done with this chapter, we will have the basic understanding of self-similar phenomena and when we are done with the whole book we will be able to appreciate the fact that self-similar phenomena are ubiquitous in nature as it is virtually everywhere we look. In our endeavour we need to spend just a few more minutes laying the foundation for that knowledge. We begin our exploration with three more examples of self-similarity.

Firstly, think of a lone leafless tree standing against a gray winter sky as the background. Needless to mention that branching patterns are quite familiar to us. There is a single trunk that rises to a region from which major branches split off. If we follow any one of these major branches, we see that they also split into smaller branches, which split into even smaller branches, and so on. A unit pattern (in this case one length splitting into two or more thinner branches) is repeated on ever-smaller size scales until we reach the tree-top. Having this picture in mind, we can state a simple generic definition: the fundamental principle of a self-similar structure is the repetition of a unit pattern on different size scales. Note that the delicate veins of leaves also have self-similar branching patterns, as do virtually all root systems. A tree is therefore the epitome of self-similarity.

Secondly, the decimal number system which is probably the oldest and most useful construction that used the idea of self-similarity. However, we use it so much in our daily life that we hardly had time to appreciate its self-similar properties. Let us look at the meter stick, which has marks for decimeters (ten of it make a meter), centimeters (ten of it makes a decimeter and a thousand make a meter), millimeters (ten of it makes a centimeter). In a sense a decimeter together with its marks looks like a meter with its marks, however, scaled down by a factor of 10. This is not an accident. It is in strict correspondence with the decimal system. When we say 248 *mm*, for example, we mean 2 decimeters, 4 centimeters, and 8 millimeters. In other words, the position of the figures determines their place value, exactly as in the decimal number system. One meter has a thousand millimeters to it and when we have to locate position 248 only a fool would start counting from left to right from 1 to 248. rather we would go to the 2 decimeter tick mark, from there to the four centimeter tick mark, and from there to the 8 millimeter tick mark. Most of us take this elegant procedure for granted. But somebody who has to convert miles, yards, and inches can really appreciate the beauty of this system. Actually finding a position on the meter stick corresponding to the branches of a tree, the decimal number. the structure of the tree expresses the self-similarity of the decimal system very strongly.

Thirdly, distant mountain ranges which are an example of statistical self-similarity as we find a hierarchy of peak/valley morphologies on size scales ranging from kilometers to centimeters. This is in sharp contrast to the above mentioned decimal system. Such self-similarity is found true in almost all the surfaces we encounter since the majority of them have an hierarchy of self-similar structure over at least some range of size scales. Consider the vast craters-within-craters surface of the Moon displaying self-similar divots that range in diameter from kilometers to millimeters. Erosion processes in desert regions yield a wealth of self-similar surfaces, and areas of snow and ice, from the vast continent of Antarctica to the piles of snow next to our driveways, have self-similar structures. There are examples of statistical self-similarity in the plant kingdom too. Ferns and cedar boughs have pleasing multi-leveled designs that are strongly self-similar. The overall shape of the fronds or boughs is roughly that of a broad sword with sharp tip and a stem running along its mid-line. However, a closer look reveals that it is actually composed of many small swords oriented at right angles to the main stem. These second-level swords, in turn, are divided into third-level swords lying roughly perpendicular to the second-level stems. Perhaps many of us had been looking on countless occasion at many seeming self-similar structure without really appreciating their true self-similar design.

One may even find self-similarity in social systems if one thinks about how our governments, judicial systems, law enforcement agencies or various economic systems work. It is fairly easy to see that they have almost discrete hierarchical arrangements with federal, state, local upto family levels. Basically a similar type of activity is occurring at the different scales and so we find a familiar design in a new context. The main principle of self-similarity—same thing on different scales— can be found in fact in countless examples from the social sectors. The Internet has grown according to the laws of self-similarity and its complex multi-scaled networking is actually resulting from the repetition of simple law. Multi-scaled networking is reminiscent of the self-similar networking structure in the brain of living systems. Likewise the growth patterns and inter-connectedness of cities exhibit similar phenomena on different scales, the hallmark of self-similarity. We could go on with further examples of self-similarity such as lightning bolts, designs on shells, aggregation of bacteria or metal ions, surfaces of cancer cells, crystallization patterns in agate, scores of scaling laws in biology, quantum particle paths, gamma-ray burst fluctuations, species distributions or abundances, drop formation, renormalization in quantum electrodynamics, and so on. We will have many opportunities in the chapter to come to add more examples because it can said beyond any reasonable doubt that nature adores self-similarity.

In physics we often investigate physical problems which we do not see by our naked eye and hence cannot appreciate straight away if the system possess self-similarity. However, we can attempt to solve them through modeling which is often governed by or described by kinetic or rate equation approach. Mathematician and physicists often look for a scaling or self-similar solution to their respective equations which is the solution in the long time limit. In this limit, the solution usually assumes a simpler and universal form. An equation is considered to assume self-similar or scaling solution if its solution has either power-law form or has the dynamic scaling form which are discussed below.

1.6 Dynamic scaling

There are many phenomena which physicists often investigate that are not static rather evolve probabilistically with time. Our universe is perhaps one of the best examples which is expanding ever since the Big Bang. Similarly, growth of networks like WWW, the Internet, etc. are also ever growing systems. Another example is the polymer degradation process where degradation does not occur in a blink of an eye rather it happens over quite a long time. Spread of biological and computer viruses too does not happen over night. In such systems we find certain stochastic variable x which assumes values that depend on time. In such cases, we are often interested to know the distribution of x at various instant of time, i.e., $f(x,t)$. Now the numerical value of f and the typical or mean value of x may well be very different at every different instant of measurement. The question is: What happens to the corresponding dimensionless variables? If the numerical values of the dimensional quantities are different, however, corresponding dimensionless quantities remain invariant then we can argue that the snapshot of the system at different times are similar. When this happens we conclude that the system is self-similar because the system is similar to itself at different time. The dynamic scaling is the litmus test of showing us that an evolving system exhibits such self-similarity.

Let us first consider that one of the variables, say t, is an independent parameter so that x can be expressed in terms of time t. Using the idea that the dimension of physical quantity must obey power monomial law we can write

$$x \sim t^z. \tag{1.33}$$

It implies that we can choose t^z as unit of measurement or yard-stick and quantify x in unit of t^z and the corresponding dimensionless quantity is

$$\xi = x/t^z. \tag{1.34}$$

Here, the quantity ξ is a number that tells how many t^z we need to measure x. If t is independent quantity then we can also express f in terms of t alone and hence we can write

$$f \sim t^\theta. \tag{1.35}$$

We can, therefore, define yet another dimensionless quantity

$$\Pi = f(x,t)/t^\theta = F(t^z\xi, t) \equiv \phi(t, \xi), \tag{1.36}$$

where the exponent θ is fixed by the dimensional requirement $[f] = [t^\theta]$ [17, 18]. Now, the numerical value of F should remain invariant despite the unit of measurement of t which is changed by some factor since ϕ is a dimensionless quantity. It implies that F must be independent of t, i.e.,

$$F(t, \xi) = \phi(\xi), \tag{1.37}$$

and hence we write

$$f(x,t) \sim t^\theta \phi(x/t^z), \tag{1.38}$$

where the function $\phi(\xi)$ is known as the scaling function. The function $f(x,t)$ is said to obey dynamic scaling if it satisfies the above relation. We thus see that the Buckingham π-theorem can provide a systematic processing procedure to obtain the dynamic scaling form.

We thus see that the emergence of dynamic scaling is deeply rooted to the Buckingham's π-theorem. Knowing that a system exhibits dynamic scaling means significant development of our understanding about the system. That is, the distribution $f(x,t)$ at various moments of time can be obtained from one another by similarity transformation

$$x \longrightarrow \lambda^z x, \ \ t \longrightarrow \lambda t, \ \ f \longrightarrow \lambda^\theta f, \tag{1.39}$$

revealing the self-similar nature of the function $f(x,t)$. One way of verifying the dynamic scaling is to plot dimensionless variables f/t^θ as a function of x/t^z of the data extracted at various different times. Then if all the plots of f vs x obtained at different times collapse onto a single universal curve then it is said that the systems at different times are similar and it obeys dynamic scaling. Essentially such systems can be termed as temporal self-similarity since the the same system is similar at different times.

1.7 Scale-invariance: Homogeneous function

A function is called scale-invariant or scale-free if it retains its form keeping all its characteristic features intact even if we change the measurement unit (scale). Mathematically, a function $f(r)$ is called scale-invariant or scale-free if it satisfies

$$f(\lambda r) = g(\lambda) f(r) \quad \forall \ \lambda, \tag{1.40}$$

where $g(\lambda)$ is a yet unspecified function. That is, one is interested in the shape of $f(\lambda r)$ for some scale factor λ which can be taken to be a length or size rescaling. For instance dimensional functions of physical quantity are always scale-free since they obey power monomial law. In fact, it can be rigorously proved that the functions that satisfy Eq. (1.40) should always have power-law form $f(r) \sim r^{-\alpha}$.

Proof: Starting from Eq. (1.40), let us first set $x = 1$ to obtain $f(\lambda) = g(\lambda) f(1)$. Thus $g(\lambda) = f(\lambda)/f(1)$ and Eq. (1.40) can be written as

$$f(\lambda x) = \frac{f(\lambda) f(x)}{f(1)}. \tag{1.41}$$

The above equation is supposed to be true for any λ, we can therefore differentiate both sides with respect to λ to yield

$$x f'(\lambda x) = \frac{f'(\lambda) f(x)}{f(1)}, \tag{1.42}$$

where f' indicates the derivative of f with respect to its argument. Now we set $\lambda = 1$ and get

$$x f'(x) = \frac{f'(1) f(x)}{f(1)} . n \tag{1.43}$$

This is a first-order ordinary differential equation which has a solution

$$f(x) = f(1) x^{-\alpha}, \tag{1.44}$$

where $\alpha = -f(1)/f'(1)$ [58]. It clearly proves that the power-law type is the only solution that can satisfy Eq. (1.40).

In fact, it can also be proved rigorously that $g(\lambda)$ too has the power-law solution.

Proof: Suppose we make two changes of scale, first by a factor of μ and then by by a factor of λ. The homogeneity condition of the function $f(r)$ implies

$$f\big(\lambda(\mu r)\big) = g(\lambda)f(\mu r) = g(\lambda)g(\mu)f(r). \tag{1.45}$$

A similar result can also be obtained by a single change of scale as follows

$$f\big((\lambda\mu)r\big) = g(\lambda\mu)f(r). \tag{1.46}$$

By applying the principle of equivalence of the above two equations we obtain

$$g(\lambda\mu) = g(\lambda)g(\mu). \tag{1.47}$$

Now any continuous function that satisfies the above functional equation must be either identically zero or else will have a simple power-law form with respect to its argument. To prove this we take the derivative with respect to μ, we find

$$\frac{\partial}{\partial\mu}g(\lambda\mu) = \lambda g'(\lambda\mu) = g(\lambda)g'(\mu), \tag{1.48}$$

where prime indicates the derivative with respect to the argument of the corresponding function. We now set $\mu = 1$ and $g'(1) = p$ then we can immediately write

$$\frac{1}{g(\lambda)}g'(\lambda) = \frac{p}{\lambda}. \tag{1.49}$$

Integrating it we find

$$\ln[g(\lambda)] = p\ln[\lambda] + c, \tag{1.50}$$

or

$$g(\lambda) = e^c\lambda^p. \tag{1.51}$$

Now from the above equation we have $g'(\lambda) = pe^c\lambda^{p-1}$ and the definition $g'(1) = p$ implies that the integration constant c has the value zero. Thus

$$g(\lambda) \sim \lambda^p, \tag{1.52}$$

and it makes the proof complete.

The power-law type distributions $f(x) \sim x^{-\alpha}$ are regarded as scale-free because $f(\lambda x)/f(x)$ depends only on λ not on x and hence their distribution requires no characteristic or typical scale. The idea is that if the unit of measurement of x is increased (decreased) by a factor of λ then the numerical value of f is decreased (increased) by a factor of $g(\lambda)$ but the overall shape of the function remains invariant. It is called scale-free also because it ensures self-similarity in the sense that the function looks the same whatever scale we look at it. One immediate consequence of the scale-invariant function $f(r)$ is that if we know its value at one point says at $r = r_0$, and we know the functional form of $g(\lambda)$, then the function $f(r)$ is known everywhere. This follows because any value of r can always be written in the form $r = \lambda r_0$, and

$$f(\lambda r_0) = g(\lambda) f(r_0). \tag{1.53}$$

The above equation says that the value $f(r)$ at any point is related to the value of $f(r_0)$ at a reference point $r = r_0$ by a simple change of scale. Of course, this change of scale is, in general, not linear.

1.7.1 Scale invariance: Generalized homogeneous functions

We shall see latter in the static scaling hypothesis where we shall use the fact that the thermodynamic potentials are homogeneous of the form A function $f(x, y)$ of two independent variables x and y is said to be a generalized homogeneous function if, for all values of the parameter λ $f(x, y)$ satisfies

$$f(\lambda^a x, \lambda^b y) = \lambda f(x, y), \tag{1.54}$$

where a and b are arbitrary numbers [5, 10]. In analogy with the homogeneous function why doesn't one define the generalized homogeneous function as $f(\lambda x, \lambda y) = \lambda^p f(x, y)$? The answer is the following. It is worth noting that that Eq. (1.54) cannot be further generalized to an equation of the form

$$f(\lambda^a x, \lambda^b y) = \lambda^p f(x, y), \tag{1.55}$$

because without loss of generality we can choose $p = 1$ in the above equation; i.e., a function $f(x, y)$ that satisfies Eq. (1.54) also satisfies

$$f(\lambda^{a/p} x, \lambda^{b/p} y) = \lambda f(x, y), \tag{1.56}$$

and the converse statement is also valid. Since the above equation is of the form of Eq. (1.54), we conclude that Eq. (1.55) is no more general than Eq. (1.54). Other equivalent forms of Eq. (1.54) that frequently appear in the literature on scaling laws are

$$f(\lambda x, \lambda^b y) = \lambda^p f(x, y) \tag{1.57}$$

and

$$f(\lambda^a x, \lambda y) = \lambda^p f(x, y). \tag{1.58}$$

The main point to note here is that there are at least two undetermined parameters a and b for a generalized homogeneous function. So the litmus test for the generalized homogeneous function is given by Eq. (1.54). Say, that it is true also for any λ so that we can choose $\lambda^a = 1/x$. Then Eq. (1.54) becomes

$$f(1, y/x^{b/a}) = x^{-p/a} f(x, y), \qquad (1.59)$$

or

$$f(x, y) = x^{p/a} f(y/x^{b/a}). \qquad (1.60)$$

It implies that the two variables x and y of the function f is combined into a single term $\frac{y}{x^{b/a}}$ in a non-trivial way. Such simplification has far reaching consequence as it has been the primary essence of Widom scaling in the theory of phase transition and critical phenomena.

1.7.2 Dimension functions are scale-invariant

Dimension function of physical quantities are always homogeneous with respect to its units of measurement since they are power-law monomial

$$f(\lambda^a P, \lambda^b Q, \lambda^c R) = \lambda^\theta f(P, Q, R), \qquad (1.61)$$

where $\theta = a\alpha + b\beta + c\gamma$. Note that there exists a dimension function for every physical quantity which are always of power-law monomial type. We therefore can conclude that physical quantities are homogeneous functions which satisfy Eq. (1.54) and hence have a power-law type solution. This is the reason why power-law distribution is so ubiquitous in nature. The most interesting property of power-law which makes them interesting is their scale-free character since it makes the function look the same whatever the scale we look at it by.

1.8 Power-law distribution

Many things we measure have a typical size or scale in the sense that a typical value around which measurements are centered. The heights of human beings perhaps is the simplest example. Most adult human beings are about 1.8m tall. Of course, the actual value will vary according to the geographic location of the population. However, the extent of variation may also depend on sex. Nonetheless, the crucial point is that we neither see individuals as small as 10cm nor we see individuals as tall as 5.0m. If one plots the heights of adult person of any country or city one will definitely find that indeed the distribution is relatively narrow and peaked around a characteristic value. We can cite another example of the emergence of a typical scale: the blood pressure of the adult human population, the speeds of cars in miles per hour on the motorway. For instance, the histogram of speeds is strongly peaked around 75 *mph* in this case. Of course the exact peak value may vary depending on the quality of city, the traffic situation, etc. However, one thing that remains universal is that there is always a sharp peak around the specific value which is the point we want to make here. Also the blood pressure of normal human beings peaks around a typical or characteristic value that we usually regard as the normal pressure.

But, not all things we measure are peaked around a typical value. Some vary over several orders of magnitude. A classic example of this type of behaviour is the sizes of towns, cities or even asset size of the people of these town and cities. The largest population of any city in the US is 8.00 million for New York City as of the most recent (2000) census. The town with smallest population is harder to pin down, since it depends on what we call a town. America's smallest town is Duffield, Virginia, with a population of 52. Whichever way you look at it, the ratio of largest to smallest population is at least 150, 000. Clearly, this is quite different from what we see for heights of people.

This observation seems first to have been observed by Auerbach, however it is often attributed to Zipf. What does it mean? Let $p(x)dx$ be the fraction of cities with population between x and $x + dx$ then exhibits power-law

$$p(x) \sim x^{-\alpha}, \tag{1.62}$$

where the exponent α is of special interest. Distribution of the form (1.62) are said to follow a power-law.

1.8.1 Examples of power-law distributions

Power-law distributions occur in an extraordinarily diverse range of phenomena [58]. Some of the examples are given below.

1.8.1.1 Euclidean geometry

How do we measure the size of a set of points that may constitute a line, a square or a cube? Consider that we are to measure the size of the length L of a straight line. To quantify its size we need an yardstick, say of size δ for instance, and use it to find N needed to cover the length L. Note that δ here is an arbitrarily chosen size and hence the number

N should depend on the size of the yard stick, i.e.,

$$N(\delta) = L/\delta, \tag{1.63}$$

is an integer if a suitable size δ is chosen. Now, if we measure the same length L with a smaller size of the yard-stick $\delta' = \delta/n$ (where $n > 0$ is an integer number) then we will have a new number

$$N(\delta/n) = \frac{L}{\delta/n}. \tag{1.64}$$

Combining the above two equations (1.63) and (1.64) one can immediately find that

$$N(\delta/n) = nN(\delta). \tag{1.65}$$

Extending the idea for the case of a plane and for an object that occupies space we can write the generalized relation

$$N(\delta/n) = n^d N(\delta), \tag{1.66}$$

where $d = 1, 2, 3$, i.e., assumes the only integer number corresponding to Euclidean geometry. It implies that if the size of the yardstick is decreased by a factor of n then the numerical value of the number N is increased by a factor of n^d. It can be rigorously proved that Eq. (1.66) can only have none but the inverse power-law solution

$$N(\delta) \sim \delta^{-d}, \tag{1.67}$$

where $d = 1, 2, 3$. We can take it as the definition of dimension which provides a systematic processing procedures to find dimension of an object. For instance, if we have a regular object that occupies a plane and we want to find its dimension then the idea is as follows. We can measure it with an yard-stick of area $\delta \times \delta$ to obtain $N(\delta) = A/\delta^2$. We can then measure it with an yard-stick of area $\delta/n \times \delta/n$ and collect a data for N versus δ by varying the n value. Then, the plots of $\ln(N)$ versus $\ln(\delta)$ will always be a straight line according to Eq. (1.67) and the slope of the line will be $d = 2$ which is the dimension of the plane. Finding the dimension following this procedure is as well-known as box counting method in physics and mathematics. The power-law solution is the signature of the fact that the object in question is self-similar.

1.8.1.2 *First return probability*

Random walk problem can be best understood using the walk in one dimension which can be defined as follows: A walker walks along, say in $1D$ lattice for simplicity, and before each step, the walker flips an honest coin. If it is head, the walker takes a step forward (or to the right) and if it is tail then the walker takes a step backward (or to the left). The coin is unbiased so that the chances of getting heads or tails are equal. Often random RW is compared with a drunkard walk who is so very drunk that he may move forward or backward with equal probability and that each step is independent of the steps which are already taken. The resulting walk is so irregular that one can predict nothing with certainty about the next step. Instead, all we can talk about is the probability of his covering a specific distance in a given time and that too in the statistical sense.

Another interesting question one may ask in the context of RW problem is the following: How long does it take for a random walker to reach a given point for the first time? This is an well-known problem which is more generally known as the first passage probability. It is defined as the probability $f(r, t)$ in which the walker returns for the first time to the position r at time t [52]. We, however, will focus on the special case where the walker starts walking from the origin and we want to know the probability that the walker returns to zero for the first time after time t. This is also known as the gamblers ruin process. The statement of the gambler's ruin problem and its relation with the first return probability is trivially simple. Say, a gambler has i dollars out of total n. Say, the rule is set so that if a coin is flipped and it turns out to be head then the gambler will gain a dollar or else lose a dollar. The game continues until the gambler goes broke. However, we will restrict our discussion to only the first return probability.

Consider that we have N number of independent walkers and initially all of them are at the origin. Then all of them are set to walk at the same time and we record the time they took to return to the origin. Equivalently, one could also perform the same experiment with a single walker and let the walker return to the origin N times and we take a record of the time it took each time the walker returned. The records for the corresponding return time will have exactly the same feature as that of for N walker. If one now looks at the record they will definitely find no order which may lead to the conclusion that there cannot exist any law to characterize the emergent behaviour. However, if the size of the record is large enough by letting $N \to \infty$ and if they undergo a systematic statistical processing then it may lead to finding some order even in this seemingly disordered data records.

We now discuss the processing procedure to extract the first return probability using the records obtained from N independent walks till they return to the origin. We first bin the record into equal size say of width 0.1. That is, say the first bin goes 0 to 0.1, the second from 0.1

to 0.2 and so forth. We then find what fraction of the total N walks return to zero within these bins and which is actually the frequency of returns within a specific bin. The plot of these fractions against bin size is equivalent to plotting normal histogram of the record produced by binning them into bins of equal size of width 0.1. To check if the corresponding distribution reveals power-law or not it is better to plot the histogram on logarithmic scale. Doing exactly this with the current records gives the straight line which is the characteristic feature of the power-law distribution when plotted on the logarithmic scale. Besides, we find the slope of the straight line is exactly $3/2$ which is essentially the exponent of the power-law. However, one should also notice that the right hand end of the distribution is quite noisy reflecting scarce data points near the tail meaning each bin near the right hand end only has a few samples in it, if any at all. Thus the fractional fluctuations in the bin counts are large which appears ultimately as a fat tail.

We can solve the problem analytically. Let us assume that S_n represents the position of the walker at the end of n seconds. We say that a return to the origin has occurred at time $t = n$, if $S_n = 0$. We note that this can only occur if n is an even integer. In order to calculate the probability of a return to origin at time $2m$, we only need to count the number of paths of length $2m$ which begins and ends at the origin. The number of such a path is clearly $\begin{pmatrix} 2n \\ n \end{pmatrix}$. We first consider u_{2m} the unconstrained probability of a return to the origin at time $t = 2m$ in which the walk is allowed to return to zero as many times as it likes, before returning there again at time $t = 2m$. The probability of return to the origin at time $2n$ is therefore given by

$$u_{2n} = \begin{pmatrix} 2n \\ n \end{pmatrix} 2^{-2n}, \tag{1.68}$$

since each path has the probability 2^{-2n}. Note that there are now $u_{2n}2^{2n}$ paths of length $2n$ which have endpoints $(0,0)$ and $(2n,0)$. We make one obvious assumption that $u_0 = 1$ since the walker starts at zero.

A random walker is said to have a first return to the origin at time $2m$ if $m > 0$ and $S_{2k} \neq 0$ for all $k < m$. We define f_{2m} as the first return probability at time $t = 2m$. In analogy with u_{2n} we can also think of the expression $f_{2m}2^{2m}$ as the number of paths of length $2m$ between the points $(0,0)$ and $(2m,0)$ so that the displacement S_{2n} versus time (i.e., n) curve do not touch the horizontal axis except at the end points $(0,0)$ and $(2m,0)$. Using this idea it is easy to establish a relation between unconstrained return probability u_{2n} and the first return probability f_{2m}. There are $u_{2n}2^{2n}$ paths of length $2n$ which have end points $(0,0)$ and $(2n,0)$. The collection of such paths can be partitioned into n sets,

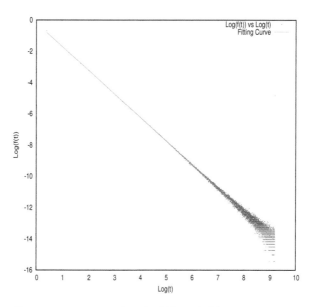

Figure 1.3: First return probability $\ln f(t)$ vs $\ln(t)$ is drawn where 25 independent realizations are superimposed. The straight line with slope equal to 1.5 clearly confirms analytical prediction.

depending upon the time of first return to the origin. Assume that a path in this collection which has a first return to the origin at time $2m$ consists of an initial segment from $(0,0)$ to $(2m,0)$, in which the path has not intercepted the time axis, and a terminal segment from $(2m,0)$ to $(2n,0)$, with no further restrictions on this segment. Thus, the number of paths in the collection which have a first return to the origin at time $2m$ is

$$f_{2m}2^{2m} \times u_{2n-2m}2^{2n-2m} = f_{2m}u_{2n-2m}2^{2n}. \qquad (1.69)$$

If we sum the above equation over m we obtain the number of paths $u_{2n}2^{2n}$ of length $2n$ and hence we have

$$u_{2n}2^{2n} = \sum_{m}^{n} f_{2m}u_{2n-2m}2^{2n}, \qquad (1.70)$$

which can be re-written as

$$u_{2n} = \begin{cases} 1 & ; \quad \text{if} \quad n = 0 \\ \sum_{m=1}^{n} f_{2m}u_{2n-2m} & ; \quad \text{if} \quad n \geq 1, \end{cases} \qquad (1.71)$$

where m is also an integer and we define $f_0 = 0$ since finding the walker at origin after 0 step is not really a return.

We find it convenient using the generating functions approach to solve the above equation for f_{2n} and hence we define

$$U(z) = \sum_{n=0}^{\infty} u_{2n} z^n, \qquad\qquad F(z) = \sum_{n=1}^{\infty} f_{2n} z^n. \quad (1.72)$$

Then multiplying Eq. (1.71) throughout by z^n and summing over whole range yield

$$
\begin{aligned}
U(z) &= 1 + \sum_{n=1}^{\infty} \sum_{m=1}^{n} f_{2n} u_{2n-2m} z^n \qquad\qquad (1.73) \\
&= 1 + \sum_{m=1}^{\infty} f_{2m} z^m \sum_{n=m}^{\infty} u_{2n-2m} z^{n-m} \\
&= 1 + F(z)U(z).
\end{aligned}
$$

So that we can immediately obtain

$$F(z) = 1 - \frac{1}{U(z)}. \quad (1.74)$$

Now, if we can find a closed form solution for the function $U(z)$, we will also be able to find closed-form solution for $F(z)$. The function $U(z)$ however is quite easy to calculate. The probability u_t that the walker is at position zero after t steps can be obtained from

$$P(m, N) = \frac{N!}{(\frac{N+m}{2})!(\frac{N-m}{2})!} \left(\frac{1}{2}\right)^N. \quad (1.75)$$

Here, $P(m, N)$ is the probability that the walker is at position m after N random steps. By setting $m = 0$ and $N = 2n$, since we need an even number of steps to reach to origin, in Eq. (1.75) we obtain

$$u_{2n} = 2^{-2n} \binom{2n}{n}. \quad (1.76)$$

so

$$U(z) = \sum_{n=0}^{\infty} \binom{2n}{n} \frac{z^n}{4^n} = \frac{1}{\sqrt{1-z}}. \quad (1.77)$$

Therefore, we have

$$F(z) = 1 - \sqrt{1-z}. \quad (1.78)$$

We find it worthwhile to note that

$$F' = \frac{U(z)}{2}, \quad (1.79)$$

where prime on the function F indicates differentiation with respect to its argument. Using Eq. (1.76) in the definition of $U(z)$ we obtain

$$U(z) = \sum_{0}^{\infty} 2^{-2n} \binom{2n}{n} z^n. \tag{1.80}$$

Substituting this into Eq. (1.79) and integrating with respect to z we can immediately obtain

$$
\begin{aligned}
f_{2n} &= \frac{u_{2m-2}}{2m} \\
&= \frac{\binom{2m-2}{m-1}}{m 2^{2m-1}} \\
&= \frac{\binom{2m}{m}}{(2m-1)2^{2m}},
\end{aligned} \tag{1.81}
$$

and we then have our solution for the distribution of first return times.

Now consider the form of f_{2n} for large n. Writing out the binomial coefficient as $\binom{2n}{n} = (2n)!/(n!)^2$, we take the logs to obtain

$$\ln f_{2n} = \ln(2n)! - 2\ln n! - 2n\ln 2 - \ln(2n-1), \tag{1.82}$$

and use the Sterling's formula $\ln n! \approx n\ln n - n + \frac{1}{2}\ln n$ to get $f_{2n} \approx \frac{1}{2}\ln n - \ln(2n-1)$ or

$$f_{2n} \approx \sqrt{\frac{2}{n(2n-1)^2}}. \tag{1.83}$$

In the limit $n \longrightarrow \infty$, this implies that $f_{2n} \sim n^{-3/2}$, or

$$f_t \sim t^{-3/2}. \tag{1.84}$$

This shows that the distribution of return times falls off as an inverse power of time t called power-law decay $f_t \sim t^{-\alpha}$ characterized by the exponent $\alpha = 3/2$.

1.8.2 *Extensive numerical simulation to verify power-law first return probability*

Consider that we have N number of independent walkers and initially all of them are at the origin. Then all of them are set to walk at the same time we then record the time they need to return to the origin where they started the walk from. Equivalently, one could also perform the same experiment with a single walker and let the walker return to the origin N times. The records for the corresponding return times will have exactly the same feature as that of for N walker. If one now looks at the record they will definitely find no order which may lead to the conclusion that there cannot exist any law to characterize the emergent behaviour. However, if the size of the record is large enough by letting N large and if they undergo a systematic statistical processing then it may lead to finding some order even in this seemingly disordered data records.

We now discuss the processing procedure to extract the first return probability using the records obtained from N independent walks till they return to the origin. We first bin the record into equal size say of width 0.1. That is, the first bin goes 0 to 0.1, the second from 0.1 to 0.2 and so forth. We then find what fraction of the total N walks return to zero within these bins. The plot of these fractions against bin size is equivalent to plotting normal histogram of the record produced by binning them into bins of equal size of width 0.1. To check if the corresponding distribution reveals power-law or not it is better to plot the histogram on a logarithmic scale. Doing exactly this with the current records gives the straight line which is the characteristic feature of the power-law distribution when plotted on the logarithmic scale. However, it is worth mentioning that the right hand end of the distribution is quite noisy reflecting scarce data points near the tail meaning each bin near the right hand end only has a few samples in it, if any at all. Thus the fractional fluctuations in the bin counts are large which appears ultimately as a fat tail.

• Zipf law: In the 1930's Zipf (the Harvard Linguistic professor) found the frequency of occurrence of some event f, as a function rank r, where the rank is determined by the extent of occurrence or frequency f, follows a power-law $f \sim r^{-k}$ with exponent close to unity. In particular, he found that the frequency of words in the English text follows such power-law. Similar distributions are seen for words in other languages too.

• Power-law has also been observed in earthquakes as observed by Guttenberg and Richter in 1954. According to Guttenberg and Richter the number of $N(E)$ of earthquakes which releases a certain amount of energy E has the form $N(E) \sim E^{-\alpha}$ with $\alpha = 1.5$ independent of geographic region. Such power-law implies that it is meaningless to define a typical or average strength of an earthquake which would correspond to the peak of a Gaussian distribution. It also implies that the same mechanism is responsible for earthquakes of all sizes, including the larges one. The power-law distribution suggests that although the large earthquakes are rare in occurrence, they are expected to occur occasionally and do not require any special mechanism. It may oppose our physical intuition, in the sense that we are used to think that small disturbances lead to small consequences and some special mechanism is required to produce large effects. It has been found that there exist a power-law relation between the amplitude (strength) and the frequency of occurrence.

In fact power-law distribution is found in an extraordinarily diverse range of phenomena that includes solar flares, computer files and wars, the numbers of papers scientists write, the number of citations received by papers, the number of hits on web pages, the sales of books, music recordings and almost every other branded commodity, the numbers of species in biological taxa, etc. just to name a few.

Chapter 2

Fractals

2.1 Introduction

Benoit B. Mandelbrot (20 November 1924–14 October 2010) born in Poland, and brought up in France and who later lived and worked in America added a new word in scientific vocabulary through his creative work which he himself coined as fractal [8, 13]. He was Sterling Professor of Mathematical Sciences at Yale University and IBM Fellow Emeritus in Physics at the IBM T.J. Watson Research Center. He was a maverick mathematician who conceived, enriched and at the same time popularized the idea of fractal almost single handedly. With the idea of fractal he has revolutionized the notion of geometry that has generated a widespread interest in almost every branch of science. He presented his new concept through his monumental book *'The Fractal Geometry of Nature'* in an awe-inspiring way and ever since then it has remained as the standard reference book for both the beginner and researcher. The advent in recent years of inexpensive computer power and graphics has led to the study of nontraditional geometric objects in many fields of science and the idea of fractals has been used to describe them. In a sense, the idea of fractal has brought many seemingly unrelated subjects under one umbrella. Mandelbrot himself has written a large number of scientific papers dealing with the fractal geometry of things as diverse as the price changes and salary distributions, turbulance, statistics of error in telephone message, word frequencies in written texts, in aggregation and fragmentation processes are just to name a few. To make his technical papers more accessible to scientific community he later wrote two more books on fractals which have inspired many to use fractal geometry in their own fields of research.

In fact, the history of describing natural objects using geometry is as old as the advent of science itself. Traditionally lines, squares, rectangles, circles, spheres, etc. have been the basis of our intuitive understanding of the geometry. However, nature is not restricted to such Euclidean objects which are only characterized typically by integer dimensions. Yet, we confined ourselves for so long to only integer dimensions. Even now in the early stage of our education we learn that objects which have only length are one dimensional, objects with length and width are two dimensional and those have length, width and breadth are three dimensional. Did we ever question why we jump to only integer dimensions? The answer to the best of my knowledge is: Never. We never questioned, if there existed objects with non-integer dimensions. However, one person did and he was none other than Mandelbrot. He was bewildered with such thoughts that for many years. He realized that nature is not restricted to Euclidean or integer dimensional space. Instead, most of the natural objects we see around us are so complex in shape that conventional Euclidean or integer dimension is not sufficient to describe them. The idea of fractal geometry appears to be indispensible for characterizing such complex objects at least quantitatively.

The idea of fractals in fact enables us to see a certain symmetry and order even in an otherwise seemingly disordered and complex system. The importance of the discovery of fractals can hardly be exaggerated. Since its discovery there has been a surge of research activities in using this powerful concept in almost every branch of scientific disciplines to gain deep insights into many unresolved problems. Yet, there is no neat and complete definition of fractal. Possibly the simplest way to define a fractal is as an object which appears self-similar under varying degrees of magnification, i.e., one loosely associates a fractal with a *shape made of parts similar to the whole in some way*. There is nothing new in it. In fact, all Euclidean object also possess this self-similar property. What is new though is the following: objects which are now known as fractals were previously regarded as geometric monsters. We could not appreciate the hidden symmetry they posses. Thanks to the idea of fractals that we can now appreciate them and also quantify them by a non-integer number exponent called the fractal dimension. The numerical value of the fractal dimension can characterize the structure and the extent of stringiness or degree of remification. This definition immediately confirms the existence of scale invariance, that is, objects look the same on different scales of observation. To understand fractals, their physical origin and how they appear in nature we need to be able to model them theoretically. The present chapter is motivated by the desire from this thirst.

2.2   Euclidean geometry

In order to grasp the notion of fractals it is absolutely necessary first to know what is Eucleadian geometry. Look at man-made objects like houses, mosques, temples, churches and furniture inside, toys, etc. and check their shapes. You will be amazed to find out that most of the objects are comprised of lines, squares, rectangles, triangles, perallelograms, semi-circles, hemispheres, spheres, etc. We all know that these objects have only integer dimensions such as 1, 2 or 3. We first encountered the concept of such integer dimensions in school where we are told that objects having only length are one dimensional, having both length and width are two dimensional and objects having length, width and height are three dimensional. A more refined definition, however, is taught later in college or university where we learn that the number of linearly independent axis needed to describe points of a given object is the geometric dimension of that object. For instance, a line is one dimensional because it takes only one number to uniquely define any point on it. That one number could be the distance from the start of the line. A plane is two dimensional since in order to uniquely define any point on its surface we require two numbers. On the other hand, objects that occupy space and require three linearly independent basis vectors to describe points of the object said to have dimension 3.

The concept of dimension is not easy to grasp. It has been one of the major challenges for mathematicians to streamline its meaning and its properties. To this end, the mathematicians have ended up with some ten or more different notions of dimension, e.g., topological dimension, Hausdorff-Besicovitch dimension, similarity dimension, box-counting dimension, informamtion dimension, Euclidean dimension, fractal dimension, etc. Some of these are related in one way or another and their details can be confusing. Some of them are more useful in some situations while others are not. In the context of fractal geometry it is more than enough to restrict ourselves to an elementary discussion on Hausdaurf-Besicovitch dimension, box-counting dimension and similarity dimension. It is note worthy to mention at this stage that the basic notion of Hausdorff-Besicovitch dimension and the box-counting dimension are the same while the later is more experimental friendly than the former. In fact, it is the Hausdorff-Besicovich dimension which we take seriously as it provides us with better understanding of fractals than others. As a passing note, we would like to mention that in

Qunatum mechanics we learnt about *Hilbertspace* which is known as the space of infinite dimensions. Mathematically, it is just an extension of Euclidean geometry as we require infinitely many linearly independent basis vectors to specify its state vectors.

Yet another description of dimensions of the Euclidean geometry based on the observation of how the "mass" of an object varies as the size of the object is changed while preserving its shape. We will see later that it plays an important role in defining fractals as well. For instance, we can take a line segment and see what happens if its size is changed by a factor of, say, two, three or k in general. The line segment can be either set points sitting next to each other leaving no holes in between or it can be full of gaps or holes of same size to form a lattice but in either case the points are uniformly distributed. Therefore, if the linear dimension of the line is increased by a factor, say of λ, then the mass of the line is also increased by the same factor and hence mass $M \sim L$ in the case of line. We can extend this argument to a circular or spherical object. The object in question can be either compact or full of holes of same size distributed uniformly. Let the diameter of the object is increased from L to λL, the mass of the object is increased by a factor of λ^2. On the other hand the mass of the object is increased by a factor of λ^3 if the object is spherical. This relationship between dimension D_E, linear scaling L and the resulting increase in mass M can be generalised to a much known mass-length relation as

$$M = L^{D_E}. \tag{2.1}$$

This relationship which holds for Euclidean shape is just telling us mathematically what we all know from our everyday experience.

An object whose mass-length relation follows Eq. (2.1) is siad to be compact and at the same time Euclidean. In general, the mass-length relation implies that if the the linear dimension of an object is increased by a factor of L while preserving its shape, the mass of the object is increased by a factor of L^d. This mass-length relation is closely related to our intuitive idea of dimension. Note that the density of the mass element of an object is defined as

$$\rho = \frac{\text{Mass of the object}}{\text{Volume of the space occupied by the object}}. \tag{2.2}$$

Now, if the dimension of the oblect d and the dimension of the embedding space D_E coincides then the object is said to be compact since its mass density $\rho \sim M/L^d$ is

$$\rho \sim L^0 = \text{const.} \tag{2.3}$$

That is why, when we are to draw 1d, 2d or 3d lattices we decorate points ensuring the fact that the density of lattice points is constant and we can have the luxury of choosing the spacing between lattice points. Do all the objects (natural or man-made) we see around us follow this mass-length relation with D_E strictly an integer number? An answer to this question is central to this chapter.

Some of the key ideas that we have learnt here are:

- Traditionally, lines, squares, rectangles, circles, spheres, etc., have been the basis of our intuitive understanding of the geometry which correspond to integer dimensions and is known as Euclidean geometry.

- Self-similarity in Euclidean objects is far too obvious, which even a layman can appreciate.

- Are there objects in nature that cannot be described by conventional Euclidean geometry?

- The question is: why do we jump to an integer value like 0 to 1, 1 to 2 and 2 to 3? We in fact took the idea of integer dimension for granted and hence prior to the 80s we never even questioned if it was possible for an object to have non-integer dimensions, at least I did not.

2.3   Fractals

Let us ask: How long is the coast of Bay of Bengal or coast of Britain? Finding an answer to this question lies in appreciating the fact that there are curves twisted so much that their length depends on the size of the yard-stick we choose to measure them, there are surfaces that fold so wildly that they fill space and their size too depends on the size of the yard-stick we choose to measure them. Consider that we are to measure the size of the coast of the Bay of Bengal, the longest beach in the world. Obviously, the size will be larger if we use an yardstick of 1.0cm long than if we use an yardstick of 1.0m long since in the former case we can capture the finer details of the twist of the coast than with the later case. Likewise, there are curves, surfaces, and the volumes in nature can be so complex and wild in character that they used to be known as the geometric monster. However, with the advent of fractal geometry they are no longer called so. The idea of fractal geometry vis-a-vis the fractal dimension provides a systematic measure of the degree of inhomogeneity in the structure. The two key ingredients for any geometric structure to be coined as fractal are (i) self-similarity, at least in the statistical sense and (ii) the dimension of the object is less than the dimension of the space where the object is being embedded.

Prior to Mandelbrot, we always took it for granted that d can only assume an integer number and the existence of objects having non-integer geometric dimension had never come under scrutiny. Now, one may ask the following:

- Are there objects which result in non-integer exponents of the mass-length relation or the power-law relation between the number N and the yardstick size δ?

- If yes, then how do they differ from those which have integer exponents?

Assume that we have infinitely many dots and we are to decorate them to constitute an Euclidean object like a line or a plane. What do we really have to ensure to that end? The dots must be embedded in an Euclidean space making sure that the dots are uniformly distributed following a regular pattern. It does not matter whether the dots are closely or widely spaced as long as they are decorated in a regular array. The constant density keeps its signature through their seemingly apparent self-similar appearence. In such cases the relation between the number N and the yardstick δ always exhibits power-law with integer exponent coinciding with the dimension of the embedding space. Can we decorate the dots in the space so that the power-law relation between N and the corresponding yardstick δ still prevails where neither the density is constant nor the self-similarity is seemingly as apparent as it was in the case of its Euclidean counterpart? It was Mandelbrot who for the first time raised the question that no one even thought of it before. Indeed, we shall see several such cases in the sections below where dots can be decorated in Euclidean space to constitute an object such that power-law relation between N and the corresponding yardstick δ still prevails but with an non-integer exponent. Below we will discuss a few simple examples where a relation between $N(\delta)$ and δ do exhibit inverse power-law with an exponent non-integer.

The geometrical structures and properties of irregular objects are addressed by Benoit B. Mandelbrot in 1975 and he coined the term "fractal" based on the Latin word fractus, derived from the content frengere meaning to break. Mandelbrot defined a fractal as a set with non-integral Hausdorff dimension which is strictly greater than its topological dimension. This definition proved to be unsatisfactory in the sense that it does not include a number of sets that are supposed to be regarded as fractals. In fact, there is no well-defined mathematical definition for characterizing the fractal. However, Falconer suggested the following which guarantees a set to be a fractal [57].

- Fractals exhibits a fine structure at arbitrary small scales. Nearly the same at every corner, on a large scale and on a small scale. Further, no smooth part should be observed by scaling.

- It has a Hausdorff dimension that is in general strictly greater than its topological dimension.

- It is recursively defined through initiators.

- Fractal are finite copies of itself.

- In general, fractals can not be easily described by Euclidean geometry!

In general, fractal are defined by rough or fragmented geometric shapes that can be split into parts where each smaller part is a resemblance of the whole. That is, fractals can be defined through self-similar property. According to self-similar property, fractals can be characterized into two types namely random fractal and deterministic fractal. An object having an approximate or statistical self-similarity is called "random fractal" and an object having regular or exact self-similarity is called "deterministic fractal" (for further reading [11, 27, 28, 57, 64, 74]).

2.3.1 Recursive Cantor set

The best known text book example of fractals is the Cantor set. The novelty of this example lies in the simplicity of its construction. The notion of fractal and its inherent character such as self-similarity is almost always introduced to the beginner through this example. In the years 1871 – 1884 Georg Cantor invented the theory of infinite sets. In the process, Cantor constructed a set that is strictly self-similar at all scales. Upon magnifying a suitable portion of the set reveals a piece that looks like the entire set itself. At the time Cantor discovered these pathological sets, it was believed that they were the purest form of mathematical invention. Never would they find an application in natural world. But, after the invention of fractal geometry they proved to be wrong. Today, we know that many natural processes produce such self-similar objects now called fractal. Almost every book written so far on fractals begins an illustration of fractals by introducing the Cantor set simply because of its simplicity in definition and yet have all the ingredients that a fractal must have.

The best aid to the comprehension of the Cantor set fractal is an illustration of its method of construction. This is given in Fig. 2.1 for

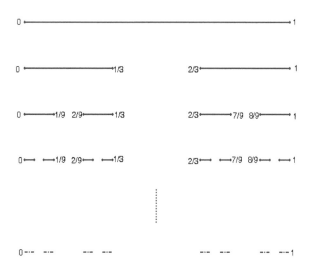

Figure 2.1: Construction of the triadic cantor set. The initiator is the unity interval $[0, 1]$. The generator removes the open middle third.

the simplest form of the Cantor set, namely the triadic Cantor set. Consider a line segment of unit length described by closed interval $C_0 = [0.1]$ which is known as an initiator. We start generating the set by diving the initiator into three equal parts and removing the middle third. We now have two closed intervals $[0, \frac{1}{3}]$ and $[\frac{2}{3}, 1]$ in step 1 each of length $\frac{1}{3}$. Thus, $C_1 = [0, \frac{1}{3}] \cup [\frac{2}{3}, 1]$. Now remove the middle thirds from the remaining two intervals leaving four closed intervals of length $\frac{1}{9}$ and these are $[0, \frac{1}{9}]$ and $[\frac{2}{9}, \frac{1}{3}]$, $[\frac{2}{3}, \frac{7}{9}]$, $[\frac{8}{9}, 1]$. This leaves $C_2 = [0, \frac{1}{9}] \cup [\frac{2}{9}, \frac{1}{3}] \cup [\frac{2}{3}, \frac{7}{9}] \cup [\frac{8}{9}, 1]$. Now remove the middle thirds from each of the remaining four intervals to create eight smaller closed intervals. Well perhaps the idea about the future steps is made clear enough. The triadic Cantor set is the limit C of the sequence C_n of sets. The sets decrease: $C_0 \supseteq C_1 \supseteq C_2 \supseteq \cdots$. So we will define the limit to be the intersection of the sets,

$$C = \bigcap_{n \in \mathbb{N}} C_n.$$

If we continue the above construction process through infinitely many steps then one may well ask: (i) What will we be left with? (ii) How much of the initiator have we thrown?

In step one we remove one interval of size $1/3$, in step two we remove two intervals of size $1/3^2$, in step three we remove four intervals of size

$1/3^3$ and in general in step n we remove 2^n intervals of size $1/3^n$. Let us add up the length of the segments we have removed:

$$1/3 + 2/9 + 4/27 + \ldots + \ldots = \frac{1}{3} \sum_{n=0}^{\infty} \left(\frac{2}{3}\right)^n \qquad (2.4)$$

$$= \frac{1}{3} \frac{1}{1 - \frac{2}{3}}$$

$$= \frac{1}{3} \times 3 = 1.$$

We thus find that asymptotically the size of the total length of all the intervals being thrown away is equal to the size of the initiator. Moreover, the set C_n consists of 2^n disjoint closed intervals, each of length $(1/3)^n$ in step n. So the total length of C_n, the sum of the lengths, is $(2/3)^n$. The limit is

$$\lim_{n \to \infty} \left(\frac{2}{3}\right)^n = 0.$$

So the total length of the Cantor set is zero. This is quite surprising as it implies that there is hardly anything left in the Cantor set, but soon we will see, there are tons of points in the Cantor set. Now let us see chek the left overs. To this end, we look into the kth moment M_k of the remaing intervals at the nth step of the construction process. The moment M_k is

$$M_k = \sum_i^{2^n} x_i^k. \qquad (2.5)$$

Note that each of the remaining intervals at the nth step are of equal size $x_i = 3^{-n}$ and hence can write

$$M_k = e^{n \ln 2 - kn \ln 3}. \qquad (2.6)$$

It means that if we choose $k = \frac{\ln 2}{\ln 3}$ then we find

$$M_{\frac{\ln 2}{\ln 3}} = 1, \qquad (2.7)$$

regardless of the value of n. That is, this result is true even in the limit $n \to \infty$. We can thus conclude that the set is not empty which is one more surprising feature of the Cantor set. We will now attempt to find the significance of the value of $k = \frac{\ln 2}{\ln 3}$.

Let us observe carefully what points form the Cantor set. We started to constitute Cantor set with the initiator $[0, 1]$ and the endpoints 0 and 1 belong to all of the future sets $C_k, k \geq n$, and therefore belong to the intersection C. Taking all the endpoints of all the intervals of all the approximations C_n, we get an infinite set of points, all belonging to C.

2.3.2 *von Koch curve*

The triadic Koch curve is another text-book like simple example used to illustrate that a curve is twisted so badly that it occupies a plane and yet has a dimension $1 < D < 2$. It is a good example of fractal because it gives a visual illustration of how simple rules can give rise to self-similar fractal structures. One of the interesting features of the Koch curve is that it is continuous but not differentiable anywhere. The construction of the Koch curve starts with the initiator which we may choose as a line segment of unit length $L(1) = 1$. It is the zeroth generation ($k = 0$) of the Koch curve. The algorithm for the construction of the Koch

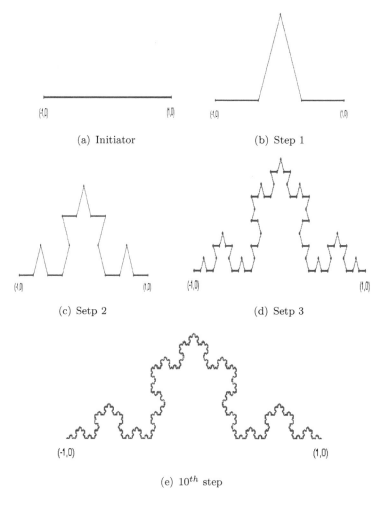

(a) Initiator   (b) Step 1

(c) Setp 2   (d) Setp 3

(e) 10th step

Figure 2.2: Construction of the *von* Koch curve.

curve is as follows: We divide the initiator into three equal pieces. The middle third is then replaced with two equal segments, both one-third in length, which form an equilateral triangle (step $k = 1$) if the middle third were not removed. This step is regarded as the generator of the Koch curve. At the next step ($k = 2$), the middle third is removed from each of the four remaining line segments and each of them is replaced by the generator. The resulting cruve will have $N = 4^2$ line segments of size $\delta = 1/3^2$ and the length of the prefractal curve $L = (4/3)^2$. This process is repeated *ad infinitum* to produce the Koch curve. Once again the self-similarity of the set is evident: each sub-segment is an exact replica of the original curve. As the number of generations increase the length of the curve clearly diverges. By applying a reduced generator to all segments of a generation of the curve a new generation is obtained. Such a curve is called *pre-fractal*. Clearly in the nth step of the generation we will have $N = 4^n$ of line segments of size $\delta = 3^{-n}$. The kth moment M_k of the interval therefore is $M_k = 4^n 3^{-nk}$ and hence $M_k = e^{n \ln 4 - nk \ln 3}$. It implies that the kth moment is always equal to the size of the initiator regardless of the value of n if we choose $k = \ln 4 / \ln 3$. We will soon see that this is once again is the fractal dimension of the Koch curve.

Let us now follow the detailed calculation of how the Hausdorff-Besicovitch dimension D can be obtained in the case of triadic Koch curve. The length of the nth prefractal is given by

$$L(\delta) = (4/3)^n. \tag{2.8}$$

The length of each of the small line segments at the nth step is

$$\delta = \frac{1}{3^n}. \tag{2.9}$$

Note that the generation number n can be expressed as

$$n = -\frac{\ln \delta}{\ln 3}. \tag{2.10}$$

We then find that the number of segments $N(n) = 4^n$ and using Eq. (2.10) we find $\ln N(\delta) = -D \ln 4$. We thus immediately find that $N(\delta)$ exhibits power-law

$$N(\delta) \sim \delta^{-\frac{\ln 4}{\ln 3}}, \tag{2.11}$$

We see once again that the exponent of the inverse power-law relation between $N(\delta)$ and δ is a non-integer. First, note that the Koch curve is embedded in a plane and hence the dimension of the embedding space is $d = 2$ which is higher than its Hausdorff-Besicovitch dimension $\frac{\ln 4}{\ln 3}$. It implies that the Koch curve is a fractal with fractal dimension $d_f = \frac{\ln 4}{\ln 3}$.

Like Cantor set, the Koch curve too can be constructed by aggregation of intervals of unit size $L(0) = 1$ and unit mass to find the mass-length relation. In step one, we can assmeble three intervals of unit size and unit mass next to each other. We replace the middle one by an equilateral triangle of sides $L(1) = 1$ and delete the base. This is known as the generator which contains four line segments and hence $M = 4$ and the linear size of the curve $L = 3$. In step two we place three line segments of length $L(2) = 3$ next to each other and replace the middle one by an equilateral triangle deleting the base again. Each of these line segments is then replaced by the generator. The system in step two thus has a mass $M = 4^2$ and the linear size of the system $L = 3^2$. In step three we again place three line segments of length $L(3) = 9$ next to each other. We then replace the line segment in the middle by an equilateral triangle and delete the base. Each line segment is then replaced by the generator to give $M = 4^3$ and the linear size of the system $L = 3^3$. We continue the process *ad infinitum*. In the nth step the system will have mass $M = 4^n$ and the linear size of the system is $L = 3^n$. Like before, we can eliminate n in favur of L to obtain

$$M \sim L^{\frac{\ln 4}{\ln 3}}. \tag{2.12}$$

Using this mass-length relation in the definition of density we again find that density of intervals in the system decreases as the linear size of the system increases like $\rho \sim L^{d_f - d}$ where $d_f = \ln 4 / \ln 3$ and $d = 2$.

2.3.3 Sierpinski gasket

We can generalize the idea of the Cantor set into higher dimensions too. For instance, in the case of Sierpinski gasket the initiator is a equilateral triangle, say S_0, and the generator divides it into four equal triangle, using lines joining the midpoints of the sides and remove the interior triangle whose vertices are the midpoints of each side of the initiator but leaves boundary of the triangle. The resultant set is $S_1 \subseteq S_0$. Now each of the remaining three triangles are subdivided into four smaller triangles with edge length $1/4$, and the three middle triangles are removed. The result is $S_2 \subseteq S1$. The generator is then applied over and over again to all the available triangles, this process gives sequence S_n of sets such that $S_0 \supseteq S_1 \supseteq S_2 \supseteq \cdots$. The limit of S_n is called the Sierpinski gasket, thus $S = \bigcap_{n \in \mathbb{N}} S_n$.

It is easy to find out that at the nth step there are 3^n triangles of side $\delta = 2^{-n}$. So the total area of S_n is $3^n \cdot (1/2^n)^2 \cdot \sqrt{3}/4$. As $n \to \infty$, S_n converges to 0. The total area of the Sierpinski gasket 0.

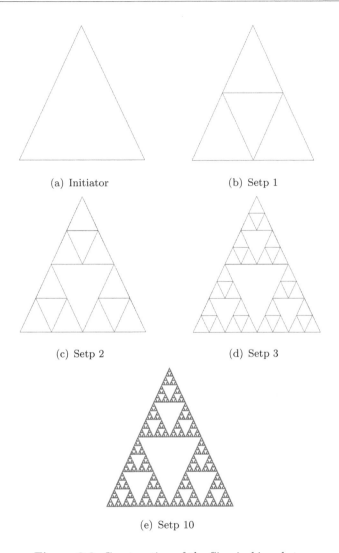

(a) Initiator

(b) Setp 1

(c) Setp 2

(d) Setp 3

(e) Setp 10

Figure 2.3: Construction of the Sierpinski gasket.

The line segments that make up the boundary of one of the triangles of S_n remain in all the later approximations $S_n, n \geq k$. So the set S contains at least all of these line segments. In S_n there are 3^n triangles, each having 3 sides of length 2^{-n}. So the total length of S is at least $3n \cdot 3 \cdot 2^{-n}$. This goes to ∞ as $n \to \infty$. So it makes sense to say that the total length of S is infinite.

At the nth step there are $N = 3^n$ triangles of side $\delta = 2^{-n}$. Eliminating n in favour of δ we find $N \sim \delta^{-\frac{\ln 3}{\ln 2}}$. One can also show that

$M \sim L^{\frac{\ln 3}{\ln 2}}$ and $M_{-\frac{\ln 3}{\ln 2}} = 1$ Likewise, in the case of Sierpinsky carpet the initiator is assumed to be a square instead of triangle and the generator divides it into b^2 equal sized squares and remove one of the square, say top left if $b = 2$ and the square in the middle if $b = 3$, is removed. This action is then applied *ad infinitum* to all the available squares in the subsequent steps. That is, in each application of the generator every survived square is divided into b^2 smaller squares and remove one in a certain prescribed fashion. The resultant fractal is called Sierpinski carpet. In the nth generation the number of squares $N = (b^2 - 1)^n$ and each squares are of equal sized with sides $\delta = (1/b)^n$. According to the definition of the Hausdorff-Besicovitch dimension we use δ as the yardstick to measure the set created in the nth generation and find that the number N scales with δ as

$$N(\delta) \sim \delta^{-\ln(b^2 - 1)/\ln b}. \tag{2.13}$$

Following the same procedure we can also obtain a similar relation for the Sierpinsky gasket

$$N(\delta) \sim \delta^{-\ln 3/\ln 2}. \tag{2.14}$$

We thus find that the dimension of the Seirpinsky gasket or the Seirpinsky carpet is less than the dimension of their embedding space $d = 2$ in both the cases the exponents are non-integer. The conservation law

$$M_{-\frac{\ln 3}{\ln 2}} = 1, \tag{2.15}$$

are obeyed here as well regardless whether the object is Sierpinsky gasket or it is the Sierpinsky carpet. So far we have looked at constructions of strictly deterministic fractals embedded in a line (Cantor set) and embedded in the plane (Koch curve and Sierpinski gasket and carpet). We can also construct fractals embedded in 3d space known as the Menger sponge.

Aforementioned attractors *von* Koch curve and Sierpinski gasket have infinite length as well as zero area. If they have been thought as 1-dimensional objects, it is too big. If they have been considered as 2-dimensional objects, it is too small. Thus, Cantor set, *von* Koch curve and Sierpinski gasket are satisfying the following:

- They can not be easily described by Euclidean geometry.

- Exhibits a fine structure at arbitrary small scales. Nearly the same at every corner, on the large scale and on a small scale. Further, no smooth part should be observed by scaling. (Self-similar property).

- It has a Hausdorff dimension that is in general strictly greater than its topological dimension. (This point will be verified in later chapter).

- It is recursively defined through initiators in the sense of iterated function system.

- Finite copies of itself. Since all are obtained from the self-referential equation $F(A) = \bigcup_{k \in N_n} f_k(A)$.

According to self-similar properties, fractals can be characterized into two types,

(i) an object having approximate or statistical self-similarity called random fractal.

(ii) an object having regular or exact self-similarity called deterministic or regular fractal.

The central theme of this book is to present the clear idea about "deterministic fractal", "random fractal" and its applications. Subsequence of this section concisely presents the mathematical backgrounds which are need to construct the deterministic fractal in the metric space. Later, we discuss the random fractals.

2.4 Space of fractal

2.4.1 Complete metric space

Definition 2.1 *[7, 12]*

A metric space (X, d) is a nonempty set X together with a function $d : X \times X \to [0, \infty)$, which is called as a metric or distance function, satisfying

(i) $d(x, y) = 0 \iff x = y$

(ii) $d(x, y) = d(y, x) \; \forall x, y \in X$

(iii) $d(x, z) \leq d(x, y) + d(y, z) \; \forall x, y, z \in X.$

The nonnegative real number $d(x, y)$ measures the distance between x and y in X. Since,

(i) Distance is zero if and only if two objects lie on the same position.

(ii) Distance is symmetry with respect to the position of object, that is the distance between two points is same whichever the point it is measured.

(iii) The distance between two points x, y cannot exceed sum of the distances from x to an intermediate point z and from y to z.

Example 2.1 *The set of all real numbers $\mathbb{R}$ together with $d(x, y) = |x - y|$ is metric space and it is denoted as $(\mathbb{R}, |.|)$, where*

$$|x| = \begin{cases} x & \text{if } x > 0, \\ -x & \text{if } x < 0, \\ 0 & \text{if } x = 0. \end{cases}$$

The distance function $|.|$ is called the usual metric on $\mathbb{R}$.

Proof: The range of modulus function $|.|$ is $[0, \infty)$, hence $d(x, y) \geq 0$ for all $x, y \in \mathbb{R}$.

(i) $|x - y| = 0 \iff x = y$

(ii) $|x - y| = |-(x - y)| = |y - x|$

(iii) Observe that $|x| \geq \pm x$ for all $x, y \in \mathbb{R}$. By using this inequality it is easy to derive the triangle inequality for $|.|$,

$$|x - y| = |x - z + z - y| \leq |x - z| + |z - y|.$$

Example 2.2 *Let X be a nonempty set. Define*

$$d(x, y) = \begin{cases} 0 & \text{if } x = y, \\ 1 & \text{if } x \neq y. \end{cases}$$

Then, d defines the metric on X and (X, d) is called the discrete metric space.

Example 2.3 *Let $X = \mathbb{R}^n$, the set of ordered n-tuples of real numbers. Define*

i. $d_1(x, y) = \sum_{k=1}^{n} |x_i - y_i|$,

ii. $d_p(x, y) = \left(\sum_{k=1}^{n} (x_i - y_i)^2 \right)^{1/2}$, $p \geq 1$,

iii. $d_\infty(x, y) = \max_{1 \leq k \leq n} \{|x_i - y_i|\}$,

for all $x, y \in \mathbb{R}^n$. Then, $d_1(x, y), d_p(x, y), d_\infty(x, y)$ are metrics on $\mathbb{R}^n$.

Definition 2.2 *A function f from a metric space (X, d_1) to another metric space (Y, d_2) and $a \in X$. We say that f is continuous at a, if for any given $\epsilon > 0$, there exists $\delta > 0$ such that*

$$d_1(x, a) < \delta \text{ implies } d_2(f(x), f(a)) < \epsilon \text{ for all } x \in X.$$

Definition 2.3 *A function $f : X \to Y$ is said to be uniformly continuous on X if for each $\epsilon > 0$, there exists $\delta > 0$ such that*

$$d_1(x, y) < \delta \text{ implies } d_2(f(x), f(y)) < \epsilon \text{ for all } x, y \in X.$$

Example 2.4 *Let $C[a, b]$ be set of all continuous real-valued functions defined on the closed interval $[a, b]$. Consider the function $|.|_\infty : C[a, b] \times C[a, b] \to [0, \infty)$ defined by*

$$|f - g|_\infty = \sup_{x \in [a,b]} |f(x) - g(x)|.$$

This function $|.|_\infty := d_\infty$ defines the metric on the set $C[a, b]$ and it is called uniform metric.

The metric d_∞ measures the distance from f to g as the maximum of vertical distance from the point $(x, f(x))$ to $(x, g(x))$ on the graphs of f and g.

Definition 2.4 *Let (X, d) be a metric space. Given a point $x \in X$ and a positive real number r. The subsets*

$$B(x, r) = \{y \in X : d(x, y) < r\}$$

and

$$B[x, r] = \{y \in X : d(x, y) \le r\}$$

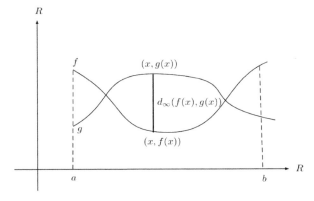

Figure 2.4: Distance between $f(x)$ and $g(x)$.

are respectively called the open and closed ball centred at x with radius r with respect to the metric d.

Definition 2.5 *Let (X, d) be a metric space and $A \subseteq X$. A point $x \in A$ is said to be an interior point of A if there exists $r > 0$ such that $B(x, r) \subseteq A$. Other words, $B(x, r) \cap A = B(x, r)$.*

A point $x \in X$ is said to be a limit point of A if for every $r > 0, B(x, r) \cap A \neq \emptyset$. Other words, each open ball centred at x contains at least one point of A different from x.

Definition 2.6 *Let (X, d) be a metric space. A sequence in X is a function f from the set of all natural numbers $\mathbb{N}$ into X. We denote the sequence as $(x_n)_{n \in \mathbb{N}}$ where x_n is image of n under the function f.*

A sequence (x_n) in X is said to be converges to the point $x \in X$ if given $\epsilon > 0$, there exists $n_0 \in \mathbb{N}$ such that $d(x_n, x) < \epsilon$ for all $n \geq n_0$. The point x is called the limit of the sequence (x_n) and we write $\lim_{n \to \infty} x_n = x$ or $x_n \to x$ as $n \to \infty$.

Definition 2.7 *A sequence $\{x_n\}_{n=1}^{\infty}$ in a metric space (X, d) is called a Cauchy sequence if for every $\epsilon > 0$, there exists $n_0 \in \mathbb{N}$ such that $d(x_n, x_m) < \epsilon$ whenever $n, m \geq n_0$.*

Theorem 2.1 *Every convergent in a metric space is a Cauchy sequence.*

Proof: Let (x_n) converge to x in a metric space (X, d). Given $\epsilon > 0$, we can choose $n_0 \in \mathbb{N}$ such that $d(x_n, x) < \epsilon/2$ for $n \geq n_0$. Then, if $n, m \geq n_0$,

$$d(x_n, x_m) \leq d(x_n, x) + d(x_m, x) < \epsilon/2 + \epsilon/2 = \epsilon.$$

Hence (x_n) is a Cauchy sequence.

Consider the metric space $(\mathbb{Q}, |.|)$ and the sequence $x_n = \frac{\lfloor n\sqrt{2} \rfloor}{n}$ in $\mathbb{Q}$. Clearly, the sequence (x_n) converges to $\sqrt{2}$. Using the Theorem 2.1, one can show (x_n) is a Cauchy sequence in $(\mathbb{Q}, |.|)$. However, it does not converge to a point of $\mathbb{Q}$. It seems that the converse of Theorem 2.1 is not true.

Proposition 2.1 *Let $\{x_n\}_{n=1}^{\infty}$ be a sequence in a metric space (X, d) and $d(x_n, x_{n+1}) < \frac{1}{2^n}$ for all n. Then $\{x_n\}_{n=1}^{\infty}$ is a Cauchy sequence.*

Proof: Let $\{x_n\}_{n=1}^{\infty}$ be a sequence in a metric space (X, d). Let $\epsilon > 0$ and choose $n_0 \in \mathbb{N}$ such that $\frac{1}{2^{n_0 - 1}} < \epsilon$. Then for all $n > m \geq n_0$ we have

$$d(x_m, x_n) \leq \sum_{k=m}^{n-1} d(x_k, x_{k+1})$$

$$< \sum_{k=m}^{n-1} \frac{1}{2^k}$$

$$< \sum_{k=m}^{\infty} \frac{1}{2^k} = \frac{1}{2^{m-1}}$$

$$\leq \frac{1}{2^{n_0-1}} < \epsilon.$$

Finally we have $d(x_m, x_n) < \epsilon$ for all m, n. Therefore $\{x_n\}_{n=1}^{\infty}$ is a Cauchy sequence.

Definition 2.8 *A metric space (X, d) is said to be complete if every Cauchy sequence in X converges to an element in X.*

2.4.2 Banach contraction mapping

Definition 2.9 *Let (X, d) be a metric space. A self-mapping f on X is said to be contraction mapping (contraction) if there exists a constant $\alpha \in [0, 1)$ such that*

$$d(f(x), f(y)) \leq \alpha d(x, y) \text{ for all } x, y \in X. \tag{2.16}$$

The constant α is called the contraction factor.

Definition 2.10 (fixed point) *Let (X, d) be a metric space and $f : X \to X$ be a mapping. A point x^* is said to be a fixed point of f if $f(x^*) = x^*$.*

Example 2.5 *Consider the space $\mathbb{R}$ with usual metric.*

1. *Let $f : \mathbb{R} \to \mathbb{R}$ be a mapping defined by $f(x) = ax$, where $a \in \mathbb{R}$. Then $x^* = 0$ is the fixed point of f*

2. *If $f : \mathbb{R} \to \mathbb{R}$ defined by $f(x) = x + a$, where $a \in \mathbb{R}$ and $a \neq 0$ then f has no fixed point.*

3. *If $f : \mathbb{R} \to \mathbb{R}$ defined by $f(x) = x$ then all the points in $\mathbb{R}$ is a fixed point of f.*

Theorem 2.2 *A contraction mapping on the metric space (X, d) is uniformly continuous function.*

Proof: Let $\epsilon > 0$ be given. If $\alpha = 0$, then $d(f(x), f(y)) = 0 < \epsilon$. If $\alpha \in (0, 1)$, then $d(f(x), f(y)) \leq \alpha d(x, y)$ for all $x, y \in X$. Choose $\delta = \epsilon/\alpha$, we have

$$d(f(x), f(y)) < \delta \text{ for all } x, y \in X.$$

This shows f is uniformly continuous on X.

The most fascinated result connecting the contraction mapping and complete metric space was explored by Stefan Banach in 1922 which stated as *every contraction mapping on a complete metric space has a unique fixed point*. Later on, this theorem is named as **Banach fixed point theorem**.

Theorem 2.3 *Let (X, d) be a complete metric space and f be a contraction mapping on X. Then f has a unique fixed point x^*.*

Proof: Let $x_0 \in X$ be an arbitrary point such that $f(x_0) = x_0$, otherwise nothing to prove. Define, $x_1 = f(x_0), x_{n+1} = f(x_n)$ for $n \in \mathbb{N}$. We claim that the sequence (x_n) is Cauchy in X. Observe,

$$\begin{aligned}
d(x_n, x_{n+1}) &= d(f(x_{n-1}), f(x_n)) \\
&\leq \alpha d(x_{n-1}, x_n) = \alpha d(f(x_{n-2}), f(x_{n-1})) \\
&\leq \alpha^2 d(x_{n-2}, x_{n-2})
\end{aligned}$$

$$\vdots$$

$$\leq \alpha^{n-1} d(f(x_1), f(x_0)) \leq \alpha^n d(x_1, x_0).$$

For any $n, m \in \mathbb{N}$ with $n < m$,

$$\begin{aligned}
d(x_m, x_n) &\leq d(x_m, x_{m-1}) + d(x_{m-1}, x_{m-2}) + \cdots + d(x_{n+1}, x_n) \\
&= \sum_{i=n}^{m-1} d(x_i, x_{i+1}) \\
&\leq \sum_{i=n}^{m-1} \alpha^i d(x_1, x_0) \leq \frac{\alpha^n}{1-\alpha} d(x_1, x_0).
\end{aligned}$$

Hence, for given $\epsilon > 0$, choose n_0 large enough that $\frac{d(x_0, x_1)\alpha^n}{1-\alpha} < \epsilon$. Then, for $n, m \geq n_0$, we have $d(x_m, x_n) < \epsilon$. It shows that (x_n) is Cauchy. Since X is complete, there exists $x^* \in X$ such that $x_n \to x^*$ as $n \to \infty$. Given f is contraction and hence f is continuous. Therefore, $x_n \to x^*$ implies $f(x_n) \to f(x^*)$. It follows $f(x^*) = x^*$.

To prove the uniqueness of fixed point x^*, assume x^*, y^* are two fixed points of f. Then $f(x^*) = x^*$ and $f(y^*) = y^*$. Since f is contraction mapping,

$$d(x^*, y^*) = d(f(x^*), f(y^*)) \leq \alpha d(x^*, y^*) < d(x^*, y^*),$$

which is contradiction. Hence $x^* = y^*$.

Note 2.1 *1. Let $X = (0,1)$ and $f : X \to X$ defined by $f(x) = \frac{x}{a}$ where $a \in \mathbb{R}$ and $a \neq 0$. Then f has no fixed point. Here X is not a complete metric space.*

2. *If $X = [a, \infty), a \geq 1$ and $f : X \to X$ defined by $f(x) = x + \frac{a}{x}$ then X is complete metric space but f is not a contraction. Since,*

$$|f(x) - f(y)| = |x - y|\left(1 - \frac{a}{xy}\right) < |x - y|, \quad \because a < xy.$$

Here $\alpha = 1$, it gives f is not a contraction. Therefore f has no fixed point in X.

2.4.3   Completeness of the fractal space

Let (x, d) be a complete metric space and $\mathcal{H}(X)$ be the set of all nonempty compact subsets of X. Now we define the distance between a point x in X and a compact subset A in $\mathcal{H}(X)$ as follows

$$d(x, A) = \inf\{d(x, a) : a \in A\}. \tag{2.17}$$

Remarks 2.1 *Here A is compact set so $d(x, a)$ exists and never takes the values less than zero for all $a \in A$. Hence, the set $\{d(x, a) : a \in A\}$ is nonnegative. Therefore, by the completeness axiom infimum of Eq. (2.17) exists.*

Suppose $d(x, A) = r$ and consider the open ball centered at x with radius r (see Fig. 2.5). No point of A lies in the interior of the open ball, since r is the infimum of distances from x to points in A. Hence, for any $r_1 > 0$ the open ball centred at x and radius $r + r_1$ contains at least one point of A.

Construct a sequence (a_n) in A such that $d(x, a_n) < r + 1/n$. Since A is compact, so there exists a subsequence (a_{n_k}) of (a_n) that converges to a' in A. Hence, $d(x, a') \leq r$. By definition of $d(x, A)$ we have $d(x, A) \leq d(x, a')$ which gives $d(x, A) = d(x, a')$. That is, if A is compact then, there exists a point a' in A such that $d(x, A) = d(x, a')$.

Now define the distance between two sets $A, B \in \mathcal{H}(X)$ as

$$d(A, B) = \sup\{d(a, B) : a \in A\}. \tag{2.18}$$

Remarks 2.2 *By the definition of supremum, there is a sequence (a_n) in A such that $d(A, B) = \lim_{n \to \infty} d(a_n, B)$. By Remark 2.1, there exists a sequence (b_n) in B such that $d(a_n, B) = d(a_n, b_n)$. Since A and B are*

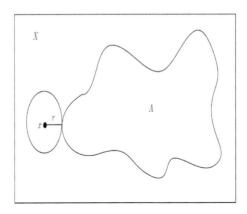

Figure 2.5: Distance from a point x to the set A.

compact, there exists subsequences (a_{n_k}) of (a_n) and (b_{n_k}) of (b_n) such that $\lim_{k \to \infty} a_{n_k} = a'$, $\lim_{k \to \infty} b_{n_k} = b'$. Therefore,

$$d(A, B) = \lim_{k \to \infty} d(a_{n_k}, B) = \lim_{k \to \infty} (a_{n_k}, b_{n_k}) = d(a', b').$$

Example 2.6 Let $A = \{x \in \mathbb{R} : x > 0\}$ and $B = \{x \in \mathbb{R} : x < 0\}$ be subsets of $\mathbb{R}$ with usual metric. Then $d(A, B) = 0$, but $A \cap B = \emptyset$. If $a = 0$ then $d(a, A) = 0$, but $a \notin A$. Thus, we observed the following:

■ $d(x, A) = 0$ does not give $x \in A$.

■ $d(A, B) = 0$ does not guarantee that A and B have common points.

Note that here A and B are not closed subsets of $\mathbb{R}$.

Exercise 2.1 Let $A, B, C \in \mathcal{H}(X)$. Then

1. Show that $d(x, A) = 0$ if and only if $x \in A$

2. Show that $d(A, B) = 0$ if and only if $A \subseteq B$

3. If $A \subseteq B$, show that $d(x, B) \leq d(x, A)$

4. $d(A \cup B, C) = \max\{d(A, C), d(B, C)\}$

Definition 2.11 Let (X, d) be a complete metric space and $\mathcal{H}(X)$ be a associated hyperspace of nonempty compact subsets of X. Then, the Hausdorff distance between A and B in $\mathcal{H}(X)$ is defined as

$$H_d(A, B) = \max\{d(A, B), d(B, A)\} \tag{2.19}$$

Theorem 2.4 *The Hausdorff distance H_d defines a metric on $\mathcal{H}(X)$.*

Proof: It is enough to verify the following conditions to say that H_d is a metric on $\mathcal{H}(X)$.

   i. $H_d(A, B) > 0$ for all $A, B \in \mathcal{H}(X)$ and $A \cap B = \emptyset$

   ii. $H_d(A, B) = 0$ if and only if $A = B$

   iii. $H_d(A, B) = H_d(B, A)$ for all $A, B \in \mathcal{H}(X)$

   iv. $H_d(A, C) \leq H_d(A, B) + H_d(B, C)$ for all $A, B, C \in \mathcal{H}(X)$

i. For any $A, B \in \mathcal{H}(X)$, by Eq. (2.17) one can get $d(A, B) \geq 0$. Let $A \cap B = \emptyset$. Then $A \not\subseteq B$ and $B \not\subseteq A$, it gives that $d(A, B) > 0$ and $d(B, A) > 0$. Thus, by the definition of H_d, the value of H_d is the maximum of two non-zero positive real numbers. Hence $H_d(A, B) > 0$.
ii. Assume $A = B$. Clearly $A \subseteq B$ and $B \subseteq A$. Therefore $d(A, B) = 0, d(B, A) = 0$ and thus $H_d(A, B) = 0$. Now assume $H_d(A, B) = 0$. This gives $d(A, B) = 0$ and $d(B, A) = 0$, since H_d is maximum of $d(A, B)$ and $d(B, A)$. But $d(A, B) = 0$ gives $A \subseteq B$ and $d(B, A) = 0$ gives $B \subseteq A$. Hence $A = B$.
iii. By the definition, H_d is the maximum of two nonnegative real numbers. We know that the maximum value is symmetric with respect to any pair of numbers, i.e., $\max\{a, b\} = \max\{b, a\}$ for any $a, b \in \mathbb{R}$. Hence $H_d(A, B) = H_d(B, A)$.
iv. Let $A, B, C \in \mathcal{H}(X)$. Then for each $a \in A$ there exists $b' \in B$ such that $d(a, B) = d(a, b')$. Now

$$
\begin{aligned}
d(a, C) &= \inf\{d(a, c) : c \in C\} \\
s &\leq \inf\{d(a, b') + d(b', c) : c \in C\} \\
&= d(a, b') + \inf\{d(b', c) : c \in C\} \\
&= d(a, B) + d(b', C).
\end{aligned}
$$

Since $a \in A$ is arbitrary and each $a \in A$ is associated with $b' \in B$ such that $d(a, B) = d(a, b')$. Take the supremum over a, then we get

$$
\sup_{a \in A} d(a, B) \leq \sup_{a \in A}\{d(a, B) + d(b', C)\}
$$
$$
d(A, C) \leq d(A, B) + d(B, C).
$$

Now, $d(A, C) \leq d(A, B) + d(B, C)$
$$
\begin{aligned}
&\leq \max\{d(A, B), d(B, A)\} + \max\{d(B, C), d(C, B)\} \\
&= H_d(A, B) + H_d(B, C).
\end{aligned}
$$

Similarly, $d(C, A) \leq H_d(C, B) + H_d(B, A)$.
Hence, $H_d(A, C) = \max\{d(A, C), d(C, A)\} \leq H_d(A, B) + H_d(B, C)$.

Definition 2.12 (Dilation) *Let $A \in \mathcal{H}(X)$ and $\epsilon \geq 0$. Then $A + \epsilon = \{x \in X : d(x, A) \leq \epsilon\}$ is called the dilation of A.*

Theorem 2.5 *For all $A \in \mathcal{H}(X)$ and $\epsilon \geq 0$, the set $A + \epsilon$ is closed.*

Proof: Let $A \in \mathcal{H}(X)$. Suppose $\epsilon = 0$ then it is clear that $A + \epsilon = A$, so $A + \epsilon$ is closed. Let $\epsilon > 0$ and x be a limit point of $A + \epsilon$. It is enough to show that $x \in A + \epsilon, i.e., d(x, A) \leq \epsilon$. As x is a limit point of $A + \epsilon$, so there exists a sequence $\{x_n\}$ in $A + \epsilon$ such that $x_n \rightarrow x$. Since $x_n \in A + \epsilon \,\forall\, n$, we get $d(x_n, A) \leq \epsilon \,\forall\, n$. For each n there exists $a_n \in A$ such that $d(x_n, A) = d(x_n, a_n)$. Thus $d(x_n, a_n) \leq \epsilon$ for all n.

The set $A \in \mathcal{H}(X)$, hence by the definition of compactness of A, every sequence $\{a_n\}$ in A has a convergent subsequence $\{a_{n_k}\}$ in A. So, the sequence $\{a_n\}$ has a subsequence $\{a_{n_k}\}$ which converges in a point of A, say a. Since, $x_n \rightarrow x$, therefore the subsequence $\{x_{n_k}\}$ of $\{x_n\}$ converges to x. Hence, $d(x_{n_k}, a_{n_k}) \rightarrow d(x, a)$. But already we have $d(x_n, a_n) \leq \epsilon$, this implies $d(x_{n_k}, a_{n_k}) \leq \epsilon$ for all k. Further, $d(x_{n_k}, a_{n_k}) \leq \epsilon$ gives that $d(x, a) \leq \epsilon$ hence $d(x, A) \leq \epsilon$. Thus $x \in A$.

Theorem 2.6 *Let $A, B \in \mathcal{H}(X)$ and $\epsilon > 0$. Then $H_d(A, B) \leq \epsilon$ if and only if $A \subset B + \epsilon$ and $B \subset A + \epsilon$.*

Proof: Suppose $H_d(A, B) \leq \epsilon$. Then $\max\{d(A, B), d(B, A)\} \leq \epsilon$, thus $d(A, B) \leq \epsilon$ and $d(B, A) \leq \epsilon$. It is enough to show that $d(A, B) \leq \epsilon \iff A \subset B + \epsilon$. Assume $d(A, B) \leq \epsilon$. Then for every $a \in A$ we have $d(a, B) \leq \epsilon$. Hence for each $a \in A$, $a \in B + \epsilon$. It gives that $A \subset B + \epsilon$. For converse part, let us assume $A \subset B + \epsilon$. Then for every $a \in A$ we get $d(a, B)\epsilon$. This implies $d(A, B) \leq \epsilon$.

Theorem 2.7 *Let $\{A_n\}_{n=1}^{\infty}$ be a Cauchy sequence in $(\mathcal{H}(X), H_d)$ and $\{n_k\}_{k=1}^{\infty}$ be a sequence of positive integer satisfies $n_k < n_{k+1}$ for all k. If $\{x_{n_k}\}_{k=1}^{\infty}$ is Cauchy sequence in X such that $x_{n_k} \in A_{n_k}$ for all k, then there exists a Cauchy sequence $\{\tilde{x}_n\}_{n=1}^{\infty}$ in X such that $\tilde{x}_n \in A_n$ for all n and $\tilde{x}_{n_k} = x_{n_k}$ for all k.*

Proof: Let x_{n_k} be a Cauchy sequence in X such that $x_{n_k} \in A_{n_k}$ for all k. For each n in between n_{k-1} and n_k choose $\tilde{x}_n \in A_n$ such that $d(x_{n_k}, A_n) = d(x_{n_k}, y_n)$. Then we get

$$d(x_{n_k}, y_n) = d(x_{n_k}, A_n) \leq d(A_{n_k}, A_n) \leq H_d(A_{n_k}, A_n).$$

Since $x_{n_k} \in A_{n_k}$, then $d(x_{n_k}, y_{n_k}) = d(x_{n_k}, A_{n_k}) = 0$. It gives $\tilde{x}_{n_k} = x_{n_k}$ for all k.

Given $\{x_{n_k}\}_{n=1}^{\infty}$ is a Cauchy sequence in X. So, for given $\epsilon > 0$, there exists a positive integer N_0 such that $d(x_{n_k}, x_j) < \epsilon/3$ for all $k, j \geq N_0$.

Since $\{A_n\}_{n=1}^{\infty}$ is Cauchy in $(\mathcal{H}(X)$, so there exists $N_1 \geq n_{N_0}$ such that $H_d(A_n, A_m) < \epsilon/3$ for all $n, m \geq N_1$. If $n, m \geq N_1$, then there exists integers $j, k \geq N_0$ such that $n_{k-1} < n \leq n_k, n_{j-1} < m \leq n_j$. Then

$$
\begin{aligned}
d(\tilde{x}_n, \tilde{x}_m) &\leq d(\tilde{x}_n, x_{n_k}) + d(x_{n_k}, x_{n_j}) + d(x_{n_j}, \tilde{x}_m) \\
&= d(x_{n_k}, A_n) + d(x_{n_k}, x_{n_j}) + d(x_{n_j}, A_m) \\
&\leq d(A_{n_k}, A_n) + d(x_{n_k}, x_{n_j}) + d(A_{n_j}, A_m) \\
&\leq H_d(A_{n_k}, A_n) + H_d(x_{n_k}, x_{n_j}) + H_d(A_{n_j}, A_m) \\
&< \epsilon/3 + \epsilon/3 + \epsilon/3 = \epsilon.
\end{aligned}
$$

Hence, $\{\tilde{x}_n\}_{n=1}^{\infty}$ is Cauchy sequence in X such that $\tilde{x}_n \in A_n$ for all n and $\tilde{x}_{n_k} = x_{n_k}$ for all k.

Theorem 2.8 *Let $\{A_n\}_{n=1}^{\infty}$ be a sequence in $(\mathcal{H}(X), H_d)$ and let A be the set of all points $x \in X$ such that there is a sequence $\{x_n\}_{n=1}^{\infty}$ that converges to x and satisfies $x_n \in A_n$ for all n. If $\{A_n\}_{n=1}^{\infty}$ is Cauchy sequence, then the set A is closed and nonempty.*

Proof: Given $\{A_n\}_{n=1}^{\infty}$ is a Cauchy sequence, therefore there exists an integer n_1 such that $H_d(A_m, A_n) < \frac{1}{2}$ for all $m, n \geq n_1$. Similarly, there exists an integer $n_2 > n_1$ such that $H_d(A_m, A_n) < \frac{1}{2^2}$ for all $m, n \geq n_2$. Continuing this process we get a sequence of integer $\{n_k\}_{k=1}^{\infty}$ such that $n_k < n_{k+1}$ for all k and $H_d(A_m, A_n) < \frac{1}{2^k}$ for all $m, n \geq n_k$. Let x_{n_1} be a fixed point in A_{n_1}. By Remark 2.1, there exists a point $x_{n_2} \in A_{n_2}$ such that $d(x_{n_1}, x_{n_2}) = d(x_{n_1}, A_{n_2})$. Then

$$
\begin{aligned}
d(x_{n_1}, x_{n_2}) = d(x_{n_1}, A_{n_2}) &\leq d(A_{n_1}, A_{n_2}) \\
&\leq H_d(A_{n_1}, A_{n_2}) < \frac{1}{2},
\end{aligned}
$$

since $\{A_{n_k}\}_{k=1}^{\infty}$ is a subsequence of a Cauchy sequence $\{A_n\}_{n=1}^{\infty}$. Similarly, there exists $x_{n_3} \in A_{n_3}$ such that

$$
\begin{aligned}
d(x_{n_2}, x_{n_3}) = d(x_{n_2}, A_{n_3}) &\leq d(A_{n_2}, A_{n_3}) \\
&\leq H_d(A_{n_2}, A_{n_3}) < \frac{1}{2^2}.
\end{aligned}
$$

Continuing this process, we can construct $\{x_{n_k}\}_{k=1}^{\infty}$ satisfying $x_{n_k} \in A_{n_k}$ and for all k,

$$
\begin{aligned}
d(x_{n_k}, x_{n_{k+1}}) = d(x_{n_k}, A_{n_{k+1}}) &\leq d(A_{n_k}, A_{n_{k+1}}) \\
&\leq H_d(A_{n_k}, A_{n_{k+1}}) < \frac{1}{2^k}.
\end{aligned}
$$

By Proposition 2.1, $\{x_{n_k}\}_{k=1}^{\infty}$ is a Cauchy sequence. Since $\{x_{n_k}\}$ is a Cauchy sequence and $x_{n_k} \in A_{n_k}$ and for all k, by Theorem 2.7, there exists a Cauchy sequence $\{\tilde{x}_n\}$ in X such that $\tilde{x}_n \in A_n$ for all n and $\tilde{x}_{n_k} = x_{n_k}$ for all k. X is complete, therefore the Cauchy sequence $\{\tilde{x}_n\}$ has a limit $\tilde{x} \in X$. For all n, $\{\tilde{x}_n\} \in A_n$ which gives $\tilde{x} \in A$. Hence A is nonempty.

Suppose s is a limit point of A. We have to prove A is closed, it is enough to prove that s belongs to A. There exists a sequence $s_n \in A\backslash\{s\}$ that converges to s, since s is a limit point of A. By the definition of A, there exists a sequence $\{t_n\}$ that converges to $\{s_n\}$ such that $t_n \in A_n$ for all n, since $\{s_n\} \subseteq A$. Moreover, there exists an integer n_1 such that $x_{n_1} \in A_{n_1}$ and $d(x_{n_1}, s_1) < 1$. Similarly, there exists an integer $n_2 > n_1$ and $x_{n_2} \in A_{n_2}$ such that $d(x_{n_2}, s_2) < \frac{1}{2}$. Continuing this way we can choose a sequence of integers $\{n_k\}$ such that $n_k < n_{k+1}$ and $d(x_{n_k}, s_k) < \frac{1}{k}$ for all k. Then

$$d(x_{n_k}, s) \leq d(x_{n_k}, s_k) + d(s_k, s)$$
$$< \frac{1}{k} + d(s_k, s) \to 0 \text{ as } k \to \infty.$$

It gives that $\{x_{n_k}\}$ converges to s. Hence, $\{x_{n_k}\}$ is a Cauchy sequence, since every convergent sequence is Cauchy sequence. It gives that $\{x_{n_k}\}$ is a Cauchy sequence for which $x_{n_k} \in A_{n_k}$ for all k. Theorem 2.7 guarantees that there exists a Cauchy sequence $\{\tilde{x}_n\}$ in X such that $\tilde{x}_n \in A_n$ for all n and $\tilde{x}_{n_k} = x_{n_k}$. Therefore $s \in A$, it shows A is closed.

Theorem 2.9 *Let $\{A_n\}_{n=1}^{\infty}$ be a sequence of totally bounded subsets of X and let A be any subset of X. If for each $\epsilon > 0$, there exists a positive integer N such that $A \subseteq A_N + \epsilon$, then A is totally bounded.*

Proof: For given $\epsilon > 0$, choose a positive integer N such that $A \subseteq A_N + \frac{\epsilon}{4}$. By the definition of totally bounded set, choose a finite set $\{x_n \in A_N : 1 \leq n \leq k\}$ such that $A_N \subseteq \bigcup_{n=1}^{k} B(x_n, \frac{\epsilon}{4})$, since A_N is totally bounded. Without loss of generality, assume that $B(x_n, \frac{\epsilon}{2}) \cap A \neq \emptyset$ for $1 \leq n \leq k$ and $B(x_n, \frac{\epsilon}{2}) \cap A = \emptyset$ for $k \leq n$. Then for each $1 \leq n \leq k$, let $y_n \in B(x_n, \frac{\epsilon}{2}) \cap A$. Let $a \in A$. To prove A is totally bounded, it is enough to prove $A \subseteq \bigcup_{n=1}^{k} B(y_n, \epsilon)$, i.e., $a \in B(y_n, \epsilon)$. Now, $a \in A$ gives $a \in A_N + \frac{\epsilon}{4}$, so $d(a, A_N) \leq \frac{\epsilon}{4}$. By Remark 2.1, there exists $x \in A_N$ such that $d(a, x) = d(a, A_N)$. Then

$$d(a, x_n) \leq d(a, x) + d(x, x_n)$$
$$\leq \frac{\epsilon}{4} + \frac{\epsilon}{4} = \frac{\epsilon}{2}.$$

Hence $x \in B(x_n, \frac{\epsilon}{2})$ for some $1 \leq n \leq k$. Thus, $y_n \in B(x_n, \frac{\epsilon}{2}) \cap A$ such that $d(x_n, y_n) < \frac{\epsilon}{2}$. It follows that

$$d(a, y_n) \leq d(a, x_n) + d(x_n, y_n)$$
$$< \frac{\epsilon}{2} + \frac{\epsilon}{2} = \epsilon.$$

Therefore for each $a \in A$ we have $y_n, 1 \leq n \leq k$, such that $a \in B(y_n, \epsilon)$, it follows that $A \subseteq \bigcup_{n=1}^{k} B(y_n, \epsilon)$. Hence, A is totally bounded subset of X.

Theorem 2.10 *Let (X, d) be a metric space and $\mathcal{H}(X)$ be a associated hyperspace of nonempty compact subsets of X with Hausdorff metric H_d. If (X, d) is complete, then $(\mathcal{H}(X), H_d)$ is complete.*

Proof: Let $\{A_n\}_{n=1}^{\infty}$ be a Cauchy sequence in $\mathcal{H}(X)$ and define A be the set of all points x in X such that there is a sequence $\{x_n\}$ that converges to x and satisfies $x_n \in A_n$ for all n. To prove the completeness of $\mathcal{H}(X)$, we need to show $\{A_n\}$ converges to A and A is a member of $\mathcal{H}(X)$.

By Theorem 2.8, the set A is nonempty closed set. It gives that A is complete, since A is closed subset of a complete metric space (X, d). Let $\epsilon > 0$. Then there exists a positive integer N such that $H_d(A_n, A_m) < \epsilon$ for all $m, n \geq$, since $\{A_n\}$ is a Cauchy sequence. Further, Theorem 2.6 gives $A_m \subseteq A_n + \epsilon$ for all $m > n \geq N$. Let $a \in A$ and fix $n \geq N$. By definition of A, there exists a sequence $\{x_k\}$ such that $x_k \in A_k$ for all k and $\{x_k\}$ converges to a. Then, Theorem 2.5 gives that $A_n + \epsilon$ is closed. Since $x_k \in A_n + \epsilon$ for each k, then it follows that $a \in A_n + \epsilon$. We took a is arbitrary element in A and showed that $a \in A_n + \epsilon$. Hence,

$$A \subseteq A_n + \epsilon. \tag{2.20}$$

By Theorem 2.9, the set A is totally bounded. Therefore A is compact, since A is nonempty, complete and totally bounded. Thus $A \in \mathcal{H}(X)$.

Let $\epsilon > 0$ and $y \in A_n$. Since $\{A_n\}$ is a Cauchy sequence, there exists a positive integer N such that $H_d(A_m, A_n) < \frac{\epsilon}{2}$ for all $m, n \geq N$ and there exists a strictly increasing sequence of positive integers $\{n_k\}$ such that $H_d(A_m, A_n) < \frac{\epsilon}{2^{k+1}}$ and $n_1 > N$ for all $m, n > n_k$. By using Remark 2.1 we get sequence $\{x_{n_k}\}$ such that $x_{n_k} \in A_{n_k}$ for all k and $d(x_{n_k}, x_{n_{k+1}}) \leq \frac{\epsilon}{2^{k+1}}$. Moreover, Proposition 2.1 guarantees $\{x_{n_k}\}$ is Cauchy sequence and Theorem 2.7 gives limit of the sequence $\{x_{n_k}\}$, a is member of A. Further,

$$d(y, x_{n_k}) \leq d(y, x_{n_1}) + \sum_{i=1}^{k} d(x_{n_i}, x_{n_{i+1}}) < \epsilon.$$

Since $d(y, x_{n_k}) \leq \epsilon$ for all k, it follows that $d(y, a) \leq \epsilon$ and hence $y \in A + \epsilon$. Thus, there exists N such that $A_n \subseteq A + \epsilon$. Moreover, from equation 2.20 $A \subseteq A_n + \epsilon$ for all $n \geq N$. Hence, by Theorem 2.6, $H_d(A_n, A) < \epsilon$ for all $n \geq N$. It seems that $\{A_n\}$ converges to A. Thus, $(\mathcal{H}(X), H_d)$ is complete.

Now we have the enough materials to present the theory of iterated function system and there by we construct the deterministic fractals.

2.5 Construction of deterministic fractals

2.5.1 *Iterated function system*

Hutchinson introduced the conventional explanation of deterministic fractals through the theory of iterated function system. Meanwhile, Barnsley formulated the theory of iterated function system called the Hutchinson-Barnsley theory in order to define and construct the fractals as a non-empty compact invariant subset of a complete metric space generated by the Banach fixed point theorem (for further reading [11, 27, 62, 86, 87, 88, 90, 91, 94, 95]). This section concisely discusses the construction of deterministic fractal (or metric fractal) in the complete metric space generated by the IFS of Banach contractions.

Definition 2.13 *For $n \in \mathbb{N}$, let $\mathbb{N}_n$ denote the subset $\{1, 2, \ldots, n\}$ of $\mathbb{N}$. Consider a finite set of contraction mappings $f_1, f_2, \ldots, f_n$ on X with contraction ratios $\alpha_k \in [0, 1), k \in \mathbb{N}_n$, simply written as $(f_k)_{k \in \mathbb{N}_n}$. Then the system $\{X; f_k : k \in \mathbb{N}_n\}$ is called an **Iterated Function System (IFS)** or finite iterated function system.*

Definition 2.14 *Define the self-mapping $F : \mathcal{H}(X) \longrightarrow \mathcal{H}(X)$ by*

$$F(A) = \bigcup_{k \in \mathbb{N}_n} f_k(A), \text{ for all } A \in \mathcal{H}(X). \tag{2.21}$$

This self-mapping F is called as Hutchinson-Barnsley mapping (HB mapping) on $\mathcal{H}(X)$.

Lemma 2.1 *Let $f : \mathcal{H}(X) \to \mathcal{H}(X)$ defined by $f(A) = \{f(a) : a \in A\}$ for all $A \in \mathcal{H}(X)$. If f is contraction on X with contraction ratio α, then f is contraction on $\mathcal{H}(X)$ with same contraction ratio α.*

Proof: Let $A, B \in \mathcal{H}(X)$. By equation 2.17, we get $d(a, B) = \min\{d(a, b) : b \in B\}$. Now, $d(f(A), f(B)) = \max\{\min\{d(f(a), f(b)) :$

$b \in B\} : a \in A\}$. Since f is contraction mapping with contraction ratio α,

$$d(f(A), f(B)) \leq \max\{\min\{\alpha.d(a, b) : b \in B\} : a \in A\}$$
$$= \alpha. \max\{\min\{d(a, b) : b \in B\} : a \in A\}$$
$$= \alpha d(A, B).$$

Similarly, $d(f(B), f(A)) \leq \alpha d(B, A)$. Hence

$$H_d(f(A), f(B)) = \max\{d(f(A), f(B)), d(f(B), f(A))\}$$
$$\leq \alpha \max\{d(A, B), d(B, A)\}$$
$$\leq \alpha H_d(A, B).$$

It shows that f is also contraction on $\mathcal{H}(X)$ with contraction ratio α.

Theorem 2.11 *Let* $A, B, C, D \in \mathcal{H}(X)$. *Then*

$$H_d(A \cup B, C \cup D) \leq \max\{H_d(A, C), H_d(B, D)\}.$$

Proof: Let $A, B, C, D \in \mathcal{H}(X)$. Then we have

$$H_d(A \cup B, C \cup D) = \max\{d(A \cup B, C \cup D), d(C \cup D, A \cup D)\}.$$

There are two possible cases i) $H_d(A \cup B, C \cup D) = d(A \cup B, C \cup D)$ and ii)$H_d(A \cup B, C \cup D) = d(C \cup D, A \cup D)$.
Case i: If $H_d(A \cup B, C \cup D) = d(A \cup B, C \cup D)$ then $d(A \cup B, C \cup D) = d(A, C \cup D)$ or $d(A \cup B, C \cup D) = d(B, C \cup D)$. Suppose $d(A \cup B, C \cup D) = d(A, C \cup D)$, then

$$H_d(A \cup B, C \cup D) = d(A, C \cup D) \leq d(A, C)$$

$$\leq H_d(A, C) \leq \max\{H_d(A, C), H_d(B, D)\}.$$

Suppose $d(A \cup B, C \cup D) = d(B, C \cup D)$, then

$$H_d(A \cup B, C \cup D) = d(B, C \cup D) \leq d(B, D)$$

$$\leq H_d(B, D) \leq \max\{H_d(A, C), H_d(B, D)\}.$$

Case ii: If $H_d(A \cup B, C \cup D) = d(C \cup D, A \cup B)$ then $d(C \cup D, A \cup B) = d(C, A \cup B)$ or $d(C \cup D, A \cup B) = d(D, A \cup B)$. Suppose $d(C \cup D, A \cup B) = d(C, A \cup B)$, then

$$H_d(A \cup B, C \cup D) = d(C, A \cup B) \leq d(C, A)$$

$$\leq H_d(A, C) \leq \max\{H_d(A, C), H_d(B, D)\}.$$

Suppose $d(C \cup D, A \cup B) = d(D, A \cup B)$, then

$$H_d(A \cup B, C \cup D) = d(D, A \cup B) \leq d(D, B)$$
$$\leq H_d(B, D) \leq \max\{H_d(A, C), H_d(B, D)\}.$$

Hence $H_d(A \cup B, C \cup D) \leq \max\{H_d(A, C), H_d(B, D)\}$.

Theorem 2.12 *Let (X, d) be a metric space and $\mathcal{H}(X)$ be a associated hyperspace of nonempty compact subsets of X with Hausdorff metric H_d. If f_k's are contraction mapping on X for all $k \in \mathbb{N}_n$, then the HB mapping F is contraction on $\mathcal{H}(X)$.*

Proof: Let $A, B \in \mathcal{H}(X)$ and consider the case $n = 2$. Then we get

$$H_d(F(A), F(B)) = H_d\left(\bigcup_{k=1}^{2} f_k(A), \bigcup_{k=1}^{2} f_k(B)\right)$$
$$\leq \max\{H_d(f_1(A), f_1(B)), H_d(f_2(A), f_2(B))\}$$
$$\leq \max\{\alpha_1 H_d(A, B), \alpha_2 H_d(A, B)\}$$
$$\leq \alpha H_d(A, B).$$

Here $\alpha = \max\{\alpha_k : k \in \mathbb{N}_2\}$.

Theorem 2.13 *Let (X, d) be a complete metric space and $(\mathcal{H}(X), H_d)$ be a associated Hausdorff metric space. If the self-mapping F, in Eq. (2.21), is defined by the IFS $\{X; f_k : k \in \mathbb{N}_n\}$, then F has a unique fixed point A^* in $\mathcal{H}(X)$, that is, there exists a unique nonempty set $A^* \in \mathcal{H}(X)$ such that F satisfying the self-referential equation*

$$A^* = F(A^*) = \bigcup_{k \in \mathbb{N}_n} f_k(A^*).$$

Moreover, for any $B \in \mathcal{H}(X)$,

$$\lim_{p \to \infty} F^{\circ p}(B) = A^*,$$

the limit being taken with respect to the Hausdorff metric.

Proof: By Theorem 2.10, completeness of the space (X, d) gives $(\mathcal{H}(X), H_d)$ is complete metric space. Theorem 2.12 shows that F is contraction mapping on $\mathcal{H}(X)$. Hence, by Banach fixed point theorem 2.3 the contraction mapping F on the complete metric space $(\mathcal{H}(X), H_d)$ has a unique fixed point. This completes the proof.

In Theorem 2.13, $F^{\circ p}$ describes the pth composition of the HB mapping F, that is, $F^{\circ p} = \underbrace{F \circ F \circ \cdots \circ F}_{p \text{ times}}$.

Definition 2.15 *A nonempty compact set A^* obtained from the Theorem 2.13 is called an invariant set or self-referential set or* **attractor** *of the IFS $\{X; f_k : k \in \mathbb{N}_n\}$.*

Note 2.2 *In general, A^* has Hausdorff dimension which exceeds its topological dimension. Hence, the fixed point $A^* \in \mathcal{H}(X)$ of the HB mapping F is called the* **Fractal** *of the IFS of Banach contractions. Sometimes A^* is called a deterministic fractal generated by the IFS of Banach contractions.*

Let us see some classical examples of deterministic fractal.

Example 2.7 (Cantor set) *Let $X = [0, 1]$ and consider the IFS on X consists of the following two mappings,*

$$f_1(x) = \frac{x}{3};$$
$$f_2(x) = \frac{x+2}{3}.$$

The self mappings f_1, f_2 are contractions with contraction factor $1/3$. By Theorem 2.13, *the Hutchinson-Barnsley mapping F of IFS $([0, 1]; f_1, f_2)$ generates the IFS and the resulting fractal is called* **Cantor set** *as shown in* Fig. 2.6.

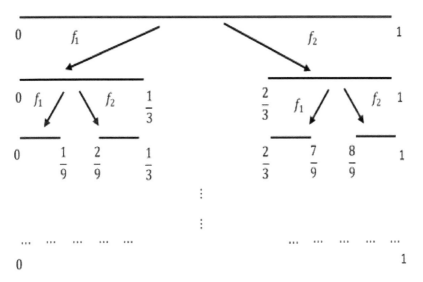

Figure 2.6: Construction of the Cantor set.

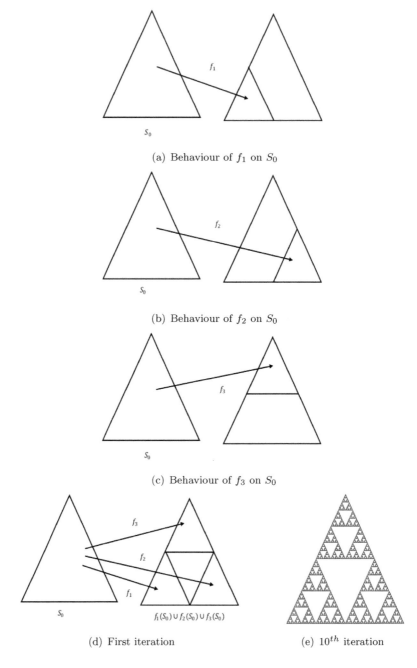

(a) Behaviour of f_1 on S_0

(b) Behaviour of f_2 on S_0

(c) Behaviour of f_3 on S_0

(d) First iteration

(e) 10th iteration

Figure 2.7: Construction of the Sierpinski gasket.

Example 2.8 (Sierpinski gasket) *Let X be an equilateral triangle $ABC \subseteq [0,1] \times [0,1]$ with the vertices $A = (0,0), B = (1,0)$ and $C = (1/2, \sqrt{3}/2)$, and consider the IFS on X consists of the following three contractions,*

$$f_1(x,y) = \left(\frac{1}{2}x, \frac{1}{2}y\right);$$

$$f_2(x,y) = \left(\frac{1}{2}x + \frac{1}{2}, \frac{1}{2}y\right);$$

$$f_3(x,y) = \left(\frac{1}{2}x + \frac{1}{4}, \frac{1}{2}y + \frac{\sqrt{3}}{4}\right).$$

*All the above contractions have contraction factor $1/2$. By Theorem 2.13, the resulting fractal is called **Sierpinski gasket** as depicted in Fig. 2.8.*

Example 2.9 (Sierpinski carpet) *Let X be a unit square in $\mathbb{R}^2$ with the vertices $A = (0,0)$, $B = (1,0)$ $C = (0,1)$ and $D = (1,1)$, and consider the IFS on X consists of the following eight contractions,*

$$f_1(x,y) = \left(\frac{1}{3}x, \frac{1}{3}y\right); f_2(x,y) = \left(\frac{1}{3}x, \frac{1}{3}y + \frac{1}{3}\right)$$

$$f_3(x,y) = \left(\frac{1}{3}x, \frac{1}{3}y + \frac{2}{3}\right); f_4(x,y) = \left(\frac{1}{3}x + \frac{1}{3}, \frac{1}{3}y\right)$$

$$f_5(x,y) = \left(\frac{1}{3}x + \frac{1}{3}, \frac{1}{3}y + \frac{2}{3}\right); f_6(x,y) = \left(\frac{1}{3}x + \frac{2}{3}, \frac{1}{3}y\right);$$

$$f_7(x,y) = \left(\frac{1}{3}x + \frac{2}{3}, \frac{1}{3}y + \frac{1}{3}\right); f_8(x,y) = \left(\frac{1}{3}x + \frac{2}{3}, \frac{1}{3}y + \frac{2}{3}\right).$$

*All the above contractions have contraction factor $1/3$. By Theorem 2.13, the resulting fractal is called **Sierpinski Carpet** as depicted in Fig. 2.8.*

Example 2.10 (von Koch curve) *Let $X = [-1,1] \times \{0\} \subseteq [-1,1]^2$ and consider the IFS on X consists of the following four contractions,*

$$f_1(x,y) = \left(\frac{1}{3}x - \frac{2}{3}, \frac{1}{3}y\right);$$

$$f_2(x,y) = \left(\frac{1}{6}x - \frac{\sqrt{3}}{6}y - \frac{1}{6}, \frac{\sqrt{3}}{6}x + \frac{1}{6}y + \frac{\sqrt{3}}{6}\right);$$

$$f_3(x,y) = \left(\frac{1}{6}x + \frac{\sqrt{3}}{6}y + \frac{1}{6}, -\frac{\sqrt{3}}{6}x + \frac{1}{6}y + \frac{\sqrt{3}}{6}\right);$$

$$f_4(x,y) = \left(\frac{1}{3}x + \frac{2}{3}, \frac{1}{3}y\right).$$

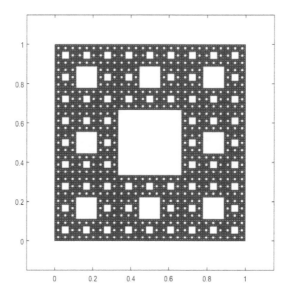

Figure 2.8: Sierpinski Carpet.

All the above contractions have contraction factor $1/3$*. By* Theorem 2.13, *the resulting fractal is called* **von Koch curve** *as shown in Fig. 2.3.2. The von Koch curve is an example of a continuous curve which is nowhere differentiable. It is also a curve of infinite length.*

Example 2.11 (Dragon curve) *Let* X *be the line segment joining two points* $(0,0), (1,0)$ *and consider the IFS on* X *consists of the following two contractions,*

$$f_1(x,y) = \left(\frac{1}{2}x - \frac{1}{2}y, \ \frac{1}{2}x + \frac{1}{2}y\right);$$

$$f_2(x,y) = \left(-\frac{1}{2}x - \frac{1}{2}y + 1, \ \frac{1}{2}x - \frac{1}{2}y\right);$$

All the above contractions have contraction factor $1/\sqrt{2}$*. By* Theorem 2.13, *the resulting fractal is called* **Dragon curve** *as shown in* Fig. 2.9.

Example 2.12 (Fractal leaves) *Let* X *be the subset of* $\mathbb{R}^2$ *and the following IFS on* X *consists of the four contractions that generate the fern leaf,*

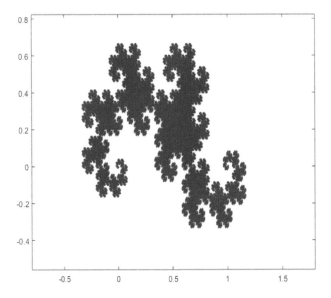

Figure 2.9: Dragon Curve.

$$f_1(x, y) = \frac{4}{25}y;$$

$$f_2(x, y) = \left(\frac{17}{20}x + \frac{1}{25}y, \ \frac{-1}{25}x + \frac{17}{20}y + \frac{4}{25} \right);$$

$$f_3(x, y) = \left(\frac{1}{2}x + \frac{-13}{50}y, \ -\frac{23}{100}x + \frac{11}{50}y + \frac{4}{25} \right);$$

$$f_4(x, y) = \left(\frac{-3}{20}x + \frac{7}{25}y, \ \frac{13}{50}x + \frac{6}{25}y + \frac{11}{25} \right).$$

The following IFS on X consists of the four contractions that generate the maple leaf,

$$f_1(x, y) = \left(\frac{49}{100}x + \frac{1}{100}y + \frac{1}{4}, \ \frac{31}{100}y + \frac{-1}{50} \right);$$

$$f_2(x, y) = \left(\frac{27}{100}x + \frac{13}{25}y, \ \frac{-2}{5}x + \frac{9}{25}y + \frac{14}{25} \right);$$

$$f_3(x, y) = \left(\frac{9}{50}x + \frac{-73}{100}y + \frac{22}{55}, \ \frac{1}{2}x + \frac{13}{50}y + \frac{2}{25} \right);$$

$$f_4(x, y) = \left(\frac{1}{25}x + \frac{-1}{100}y + \frac{13}{25}, \ \frac{1}{2}x + \frac{8}{25} \right).$$

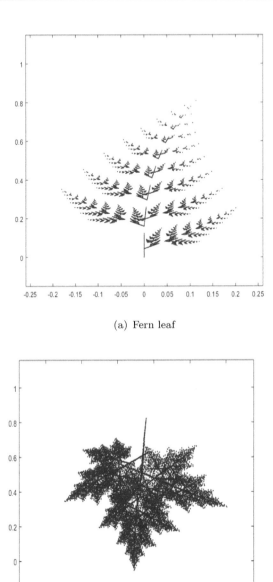

(a) Fern leaf

(b) Maple leaf

Figure 2.10: Fractal leaves.

To summarize, the deterministic fractal constructed in the example 2.7-2.12 possesses the following facts:

■ Fine structures: Nearly the same at every corner by scaling and no smooth part is observed by scaling.

■ Agrees with the Mandelbrot definition: Hausdorff dimension is strictly greater than its topological dimension.

■ A recursive construction: It is recursively defined through IFS with initiator, here initiator is line or region in $\mathbb{R}^2$ plane.

■ Self-similarity: All are finite copies of itself.

■ Due to their length and area measures, they can not be easily described by Euclidean geometry.

■ A natural pattern.

Chapter 3

Stochastic Fractal

3.1 Introduction

All the examples discussed in the previous chapter are far removed from what we have around us in nature. Nature loves randomness and the natural objects which we see around us evolve with time. The apparently complex look of most of natural objects does not mean that nature favours complextity, rather that the opposite is true. Often the inherent and the basic rule is trivially simple; it is in fact the randomness and the repetition of the same simple rule over and over again makes the object look complex. Of course, natural fractals cannot be strictly self-similar rather they are statistically self-similar [8, 13, 38]. For instance, if a child is shown a picture of a sky full of clouds distributed all over, then even the child can capture the generic feature of cloud distribution and use it in the future. Later if the same child is asked to draw the picture of clouds without looking at it, the child may well do it. The two pictures will of course never be the same but it will be similar depending on how well the child can draw. Similarly, we can all draw a curve describing the tip of the trees in the horizon but details of the two pictures drawn by the same person will never be the same despite how hard one tries. Capturing the generic feature of cloud distribution without having to know anything about self-similarity can be described as our natural instinct. In the following section we attempt to incorporate both the ingredients (randomness and kinetic) to the various classical fractals in order to know what role these two quantities play in the resulting processes [29, 35, 40, 42, 54, 55].

To this end, we will consider a couple of interesting variants of the classical Cantor set in which probability, time and randomness are incorporated in a logical progression. This allows for us to learn the role that each of these parameters plays in the resulting fractal. We first propose the dyadic Cantor set (DCS) which is simpler than the triadic Cantor set since dividing into two intervals requires only one cut while dividing into three requires two of them [85]. Note that addition of every extra cut adds extra complications in solving the problem either analytically or numerically. It is worth noting that the generator that divides an interval into two equal parts and removes one will leave nothing to investigate since there will always be only one interval left in the system. The dyadic Cantor set results in a fractal only if we remove one of the intervals with certain probability which makes it inherently a random fractal and this is sharp contrast to its triadic counterpart. We then introduce time into the problem while we still divide the intervals into two equal parts. In the kinetic DCS, we apply the generator sequentially, i.e., at each step the generator is applied to only one interval, instead of applying it recursively to all the intervals which is done in the traditional Cantor set [71]. We then further modify the generator such that it divides an interval randomly into two parts instead of dividing into two equal parts and apply it sequentially. It now incorporates both time and spatial randomness in the system making it a stochastic counterpart of the DCS. Each of these variants are solved analytically to obtain fractal dimension and to show self-similarity. Analytical results, especially the self-similar properties, are verified numerically by invoking the idea of data collapse [37].

3.2   A brief description of stochastic process

Stochastic process provides the theoretical framework for studying non-equilibrium statistical mechanics. In order to be able to understand and appreciate the meaning of a stochastic process it is essential to understand the concept of random variable as it lies at the very heart of its definition. *Stochastic or random variable x* is defined by (i) a set of possible values $x_1, x_2, .., x_n$ which we may call *range* or *set of states* and (ii) probability distribution over this set. Consider a simple example of a dice to illustrate the two points mentioned above. In each throw, the number in the upper face corresponds to the variable x, with possible outcomes: $\{1, 2, 3, 4, 5, 6\}$ and the probabilities attached with each of the outcomes which is $1/6$ (for an honest dice) in the present example. It is worthwhile to mention that the range or the set of states may be *discrete* or continuous, finite or infinite. If the set is discrete (as the case of dice, or coin) then the probability distribution will be given by a set of non negative numbers such that the attached probabilities satisfies

$$\sum_n P_n = 1. \tag{3.1}$$

On the other hand, if the range is continuous within an interval $[a, b]$ over the x axis then the probability distribution is given by a non negative function $P(x) > 0$ which is normalized in the sense that

$$\int_a^b P(x)dx = 1, \tag{3.2}$$

where the integral extends over the whole range. The probability that x has a value between x and $x + dx$ is $P(x)dx$.

Once a stochastic variable x has been defined, an infinity of other stochastic variables can be derived from it. For instance, all quantities $y = f(x)$ obtained by some mapping f is again a stochastic variable which always indicates some conceptual method. These quantities y could also be functions of an additional variable t, i.e.,

$$y(t) = f(x, t), \tag{3.3}$$

where t is the time and typically it represents the realization of the process. The function $y(t)$ describes the stochastic process if t stands for time [53, 6]. Thus a stochastic function is simply a function of two variables, one of which is the time t and the other is a stochastic variable x as defined above. In other words, systems which evolve probabilistically in time or systems in which there exists a certain time-dependent random variable $x(t)$ then it is regarded as stochastic process(refer [53, 6]).

3.3 Dyadic Cantor Set (DCS): Random fractal

Dyadic Cantor set starts with an initiator which is typically an interval of unit length $[0, 1]$. The generator then divides it into two equal parts and deletes one, say the right half, with probability $(1 - p)$. After step one, the system will have on average $(1 + p)$ number of sub-intervals each of size $1/2$, since the right half interval remains there in step one with probability p. Say that we give a piece of thread and a scissor to N number of people to divide their thread into two equal parts. We then give them a fair coin and ask them to remove one of the two pieces if the outcome of toss of the coin is head and keep both the pieces if the outcome is tail. This would correspond to $p = 1/2$ and hence in the limit $N \to \infty$ half of the people will have two intervals and the other half will have only one intervals to make the average number of intervals equal to $1 + p$ where $p = 1/2$ in this case. In the next step, the generator is applied to each of the available $(1 + p)$ sub-intervals to divide them into two equal parts and remove the right half from each of the $1 + p$ intervals with probability $(1 - p)$. The system will then have on the average $(1 + p)^2$ number of intervals of size $1/4$ as $(1 - p)(1 + p)$ number of intervals of size $1/4$ are removed on the average. The process is then continued over and over again by applying the generator on all the available intervals at each step recursively. Like its definition, finding the fractal dimension of the DCS problem is also trivially simple. According to the construction of the DCS process there are $N = (1 + p)^n$ intervals in the nth generation each of which have size $\delta = 2^{-n}$ and hence it is also the mean interval size. Like in the recursive triadic Cantor set we can construct the kth moment M_k of the intervals size at nth step of construction process of DCS too and find that here too it is a conserved quantity.

Once again the most convenient yard-stick to measure the size of the set in the nth step is the mean interval size $\delta = 2^{-n}$ which coincides with the individual size of the intervals in the nth step. Expressing N in favour of δ using $\delta = 2^{-n}$ we find that the number N falls off following power-law against mean interval size δ, i.e.,

$$N(\delta) \sim \delta^{-d_f}, \tag{3.4}$$

with $d_f = \frac{\ln(1+p)}{\ln 2} < 1 \ \forall \ 0 < p < 1$. Note that the exponent d_f is non-integer and at the same time it is less than the dimension of the space $d = 1$ where the set is embedded if we have $0 < p < 1$ then it is the fractal dimension of the resulting dyadic Cantor set [85]. Unlike triadic Cantor set where the Cantor dusts are distributed in a strictly self-similar fashion the Cantor dust in the dyadic Cantor set are distributed in a random fashion yet it is self-similar but in the statistical sense [54, 55, 66, 67, 70, 83, 85, 97].

It is well-known that the triadic Cantor set possess the following unusual property. For instance, the intervals which are removed from the set are $1/3, 2/9, 4/27, ...,$ etc. and if we add them up we get

$$\frac{1}{3}\sum_{n=0}^{\infty}(2/3)^n = 1, \tag{3.5}$$

which is the size of the initiator. This leads us to conclude that the size of the set that remains is precisely zero since the sum of the sizes that are removed equal to the size of the initiator. The question is: Does the dyadic Cantor set too possess the same properties? It is indeed the case. For instance, on the average in step one the amount of size removed is

$\frac{1-p}{2}$, in step two the total amount of size removed is $\frac{(1-p)(1+p)}{4}$, in step three it is $\frac{(1-p)(1+p)^2}{8}$, in step four it is $\frac{(1-p)(1+p)^3}{16}$ and so on. If we add these intervals we obtain

$$\frac{(1-p)}{2} \sum_{n=0}^{\infty} \left(\frac{1+p}{2}\right)^n = 1 \tag{3.6}$$

which is again the size of the initiator. It means there is hardly anything left in the set. However, we will show later that there is still tons of members in the set. One of the virtues of the dyadic Cantor set is its simplicity. The notion of random fractal and its inherent character, self-similarity, can be introduced to the beginner through this example in the simplest possible way.

3.4   Kinetic dyadic Cantor set

Now we come to the following question, which is worth addressing. What if the generator, that divides an interval into two equal parts and remove one with probability $1-p$, is applied to only one interval at each step instead of applying it to all the available intervals? Clearly, the interval sizes of the remaining intervals along the line will have great many different which is in sharp contrast to the one created by the DCS problem. It raises a further question: How do we choose one interval when the system has more than one interval of different sizes? We choose the case whereby an interval is picked with the probability proportional to their respective sizes as it appears to be the most generic case. One advantage of modifying the DCS problem in this way is that we can use the customized fragmentation equation approach to solve it analytically.

Note that the construction of Cantor set is essentially part of the fragmentation process. We can thus use the kinetics of the fragmentation equation to the dyadic Cantor set problem analytically. The kinetics of the fragmentation process can be described by the evolution of the particle (which we shall call interval) size distribution function $c(x,t)$, where $c(x,t)dx$ is the number of intervals of size within x and $x + dx$, can be described by the following integro-differnetial equation

$$\frac{\partial c(x,t)}{\partial t} = -c(x,t) \int_0^x dy F(y, x-y) dz$$

$$+ 2 \int_x^{\infty} dy c(y,t) F(x, y-x). \tag{3.7}$$

In this equation the kernel $F(x,y)$ describes the rules and rate at which a parent interval of $x+y$ is divided into two smaller intervals of size x and y [19, 21]. The first term on the right hand side of Eq. (3.7) describes the loss of interval of size x due to their division into two smaller intervals. On the other hand, the second term describes the gain of interval of size x due to division of an interval of size $y > x$ into two smaller intervals so that one of the two smaller intervals is of size x. The factor "2" in the gain term actually $\frac{2}{1}$ that describes that at each breaking event we create two intervals out of one interval.

To customize Eq. (3.7) for the kinetic dyadic Cantor set we just need to replace the factor "2" in the gain term by $1+p$ and choose the following kernel

$$F(x,y) = (x+y)\delta(x-y). \tag{3.8}$$

The delta function ensures the fact that the two product intervals x and y must be equal in size. However, it differs from the earlier definition of dyadic Cantor set Note that in the fragmentation process, at each step only one interval is divided. Thus it is necessary to choose a rule to decide how an interval is picked when there is more than one interval of different sizes. To this end, the pre-factor $(x+y)$ describes that a parent interval of size $x+y$ is chosen preferentially according to their sizes and then the delta function ensures that it is divided into two intervals of size x and y so that $x = y$. The resulting equation for kinetic dyadic Cantor set becomes

$$\frac{\partial c(x,t)}{\partial t} = -\frac{x}{2}c(x,t) + (1+p)2xc(2x,t) \tag{3.9}$$

which is the required rate equation that describes the kinetic DCS problem.

The algorithm of the jth step which starts with, say N_j number of intervals, can be described as follows.

(a) Generate a random number R from the open interval $(0,1)$.

(b) Check which of the $1, 2, ..., N_j$ intervals contains the random number R. Say, the interval that contain R is labelled as m and hence pick the interval m. Else, if none of the surviving N_j intervals contain R then increase time by one unit and go to step (a).

(c) Apply the generator to the sub-interval m to divide it into two equal pieces and remove one of the two parts with probability $(1-p)$.

(d) Label the newly created intervals starting from the left end which is labelled with its parents label m and the interval on the right if it remains there then it is labelled with a new number N_{j+1}.

(e) Increase time by one unit.

(f) Repeat the steps (a)-(e) *ad infinitum*.

To solve Eq. (3.9) we find it convenient to introduce the nth moment of $c(x,t)$

$$M_n(t) = \int_0^\infty x^n c(x,t)dx, \qquad (3.10)$$

instead of finding solution for $c(x,t)$ itself we find attempt to find the solution for its moment since finding the the latter is much simpler than the former. Incorporating it in Eq. (3.9) gives the rate equation for $M_n(t)$ which reads as

$$\frac{dM_n(t)}{dt} = -\left[\frac{1}{2} - \frac{(1+p)}{2^{n+1}}\right]M_{n+1}(t). \qquad (3.11)$$

We can easily find a value of $n = n^*$ for which M_{n^*} is a time independent or a conserved quantity simply by finding the root of the following equation

$$\frac{1}{2} - \frac{(1+p)}{2^{n^*+1}} = 0. \qquad (3.12)$$

Solving it we immediately find that $n^* = \frac{\ln(1+p)}{\ln 2}$ implying that the quantity $M_{\frac{\ln(1+p)}{\ln 2}}(t)$ is a conserved quantity. Numerically, it means that if we label all the surviving intervals, say, at the jth step as $x_1, x_2, x_3, ...x_j$ starting from the left most till the right most interval then the n^*th moment is

$$M_{n^*} = x_1^{n^*} + x_2^{n^*} + x_3^{n^*} + ... + ... + x_j^{n^*}. \qquad (3.13)$$

The numerical simulation too suggest this which is shown in Fig. (3.1). Note that the value of this moment in a given realization remains the same although the exact numerical value at which the value remains the same may vary with different realization. Interestingly, the ensemble averaged value on the other hand is equal to the size of the initiator regardless of the time we choose to measure. To find why the index of the moment $n^* = \frac{\ln(1+p)}{\ln 2}$ is so special we need to know the solution for the nth moment.

It is expected that the kinetic DCS problem too, like the DCS problem, will generate fractals in the long time limit and hence must exhibits self-similarity an essential property of fractals. It is therefore reasonable

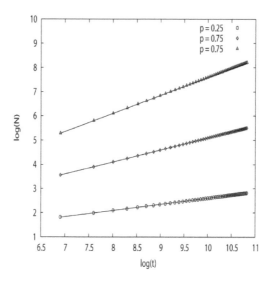

Figure 3.1: The plot show that the number of intervals N grows with time t following power-law $N(t) \sim t^{d_f}$ with exponent $n^* = \ln(1+p)/\ln 2$ which is exactly what has been predicted by Eq. (3.18).

to anticipate that the solution of Eq. (3.11) for general n will exhibit scaling. Existence of scaling means that the various moments of $c(x,t)$ should have power-law relation with time [26]. We therefore can write a tentative solution of Eq. (3.11) as

$$M_n(t) \sim A(n)t^{\alpha(n)}. \tag{3.14}$$

Substituting Eq. (3.14) in Eq. (3.11) we obtain the following recursion relation

$$\alpha(n+1) = \alpha(n) - 1. \tag{3.15}$$

Iterating it subject to the condition that $\alpha(\ln 1 + p/\ln 2) = 0$ gives

$$\alpha(n) = -(n - \frac{\ln 1 + p}{\ln 2}). \tag{3.16}$$

We therefore now have an explicit asymptotic solution for the nth moment

$$M_n(t) \sim t^{-\left(n - \frac{\ln(1+p)}{\ln 2}\right)}. \tag{3.17}$$

It implies that the number of intervals $N(t) = M_0(t)$ grows with time as

$$N(t) \sim t^{\frac{\ln(1+p)}{\ln 2}}, \tag{3.18}$$

which is verified numerically (see Fig. (3.1). On the other hand, the mean interval size $\delta = M_1(t)/M_0(t)$ decreases with time as

$$\delta \sim t^{-\gamma} \quad \text{with} \quad \gamma = 1. \tag{3.19}$$

Using it in Eq. (3.18) to eliminate time in favour of δ we find that N exhibits the a power-law

$$N \sim \delta^{-d_f}, \tag{3.20}$$

and found the same exponent $d_f = \frac{\ln(1+p)}{\ln 2}$ as n^* as the fractal dimension of DCS [23]. Besides, the value of d_f is the same as the index of the conserved moment n^*. It proves that the exact value of the fractal dimension does not depend whether we apply the generator to one interval or to all the available intervals at each step as long as the generator remain the same.

3.5   Stochastic dyadic Cantor set

Yet another interesting question is: what if we use a generator that divides an interval randomly into two smaller intervals instead of dividing

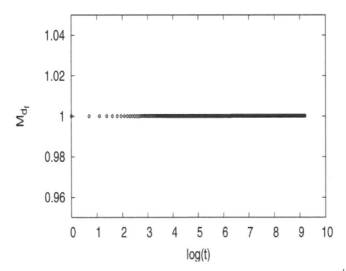

Figure 3.2: The sum of the d_fth power of all the remaining intervals $x_1^{d_f} + x_2^{d_f} + \ldots + x_j^{d_f} = 1$ regardless of time provided $d_f = \ln(1+p)/\ln 2$ which is the fractal dimension of the kinetic dyadic Cantor set. An interesting point to note that the numerical value of the conserved quantity is different in every independent realization albeit remain constant with time time.

into two equal intervals? Obviously, the interval sizes will now be a random variable x whose sample space and the corresponding probability will be different at a different time in a given realization. That is, the system will now evolve probabilistically with time in which the random variable is time dependent $x(t)$ and hence we regard the system as stochastic dyadic Cantor set. The process starts with an initiator which can be of unit interval $[0, 1]$ as before. However, the generator here divides an interval randomly into two pieces and remove one with probability $(1 - p)$. Perhaps the model can be best described by giving the algorithm for the jth generation step that starts say with N_j number of intervals. It can be described as follows.

(i) Generate a random number R from the open interval $(0, 1)$.

(ii) Check which of the available intervals contains R.

(iii) Pick the interval $[a, b]$ labelled as k if it contains R else go to step (i) if none of the available intervals contain R and increase the time by one unit in either case.

(iv) Apply the generator onto the interval k to divide it randomly into two pieces and remove one with probability $(1 - p)$. For this we generate a random number, say C, from the open interval (a, b) to divide it into $[a, c]$ and $[c, b]$ and delete the open interval (c, b) with probability $(1 - p)$.

(v) Update the logbook by labeling the left end of the two newly created interval $[a, c]$ with its parents label k since the label k was already redundant and label the other interval as $N_j + 1$ if it has not been removed in step (iv).

(vi) Repeat the steps (i)-(vi) *ad infinitum*.

The binary fragmentation equation given by Eq. (3.7) can describe the rules of the SDCS problem stated in the algorithm (i)-(vi) if we choose

$$F(x, y) = 1, \tag{3.21}$$

and the factor '2' in the gain term is replaced by $(1 + p)$. The master equation for the stochastic dyadic Cantor set then is

$$\frac{\partial c(x, t)}{\partial t} = -xc(x, t) + (1 + p) \int_x^\infty c(y, t)dy. \tag{3.22}$$

Incorporating the definition of the nth moment in it gives

$$\frac{dM_n(t)}{dt} = -\left[1 - \frac{(1 + p)}{n + 1}\right] M_{n+1}. \tag{3.23}$$

To solve it for $M_n(t)$ we follow the same procedure as for the KDCS problem and find the asymptotic solution for the nth moment as

$$M_n(t) \sim t^{(n-p)z} \quad \text{with} \quad z = -1. \tag{3.24}$$

It implies that p is the special value of n for which $M_p(t)$ is a conserved quantity. Note that once again we find that the exponent of the power-law relation for $M_n(t)$ is linear in n and hence the system must obey a simple scaling but only in the statistical sense. It is interesting to note that the nth moment $M_n(t)$ is a conserved quantity if $n = p$ whereas for KDCS the conserved quantity corresponds to $n = \ln(1+p)/\ln 2$ which is greater than p $\forall p$.

Next interesting quantity is to check if the solution for the mean interval size agrees with the numerical simulation. From Eq. (3.24) we find that the number of interval $N(t)$, which is the zeroth moment $M_0(t)$, grows with time as

$$N(t) \sim t^p, \tag{3.25}$$

and the total mass $M(t)$, which is the first moment $M_1(t)$, decreases with time as

$$M(t) \sim t^{-(1-p)}, \tag{3.26}$$

since $(1-p) > 0$ for $0 < p < 1$. Using these in the definition of the mean interval size

$$\delta = \frac{M(t)}{N(t)}, \tag{3.27}$$

we find that

$$\delta(t) \sim t^{-1}, \tag{3.28}$$

which is non-trivial as it is independent of p. Expressing the expression for the number of intervals $N(t)$ in terms δ we find it scales as

$$N(\delta) \sim \delta^{-d_f}, \tag{3.29}$$

with $d_f = p$ is the fractal dimension of the stochastic dyadic Cantor set [85]. That is the significance of p for which the pth moment is conserved according to Eq. (3.24) Note that d_f is always less than $\frac{\ln(1+p)}{\ln 2}$ $\forall 0 < p < 1$ revealing that the fractal dimension of the stochastic fractal is always less than that of its recursive or kinetic counterpart.

We still have to prove that the resulting fractal is self-similar which is one of the basic ingredients. We shall now apply the Buckingham Pi theorem to obtain scaling solution for $c(x,t)$ as it provides deep insight into the problem [37]. Note that according to Eq. (3.22) the governed parameter c for a given value of p depends on two parameters x and t.

However, the knowledge about the decay law for the mean interval size implies that one of the parameters, say x, can be expressed in terms of t since according to Eq. (3.28) t^{-1} bear the dimension of interval size x. We therefore can define a dimensionless governing parameter

$$\xi = \frac{x}{t^{-1}}, \tag{3.30}$$

and a dimensionless governed parameter

$$\Pi = \frac{c(x,t)}{t^\theta}. \tag{3.31}$$

The numerical value of the right side of the above equation remains the same even if the time t is changed by some factor, say μ for example, since the left hand side is a dimensionless quantity. It means that the two parameters x and t must combine to form a dimensionless quantity $\xi = x/t^{-1}$ and the dimensionless parameter Π can only depends on ξ. In other words we can write

$$\frac{c(x,t)}{t^\theta} = \phi(x/t^{-1}), \tag{3.32}$$

which leads to the following dynamic scaling form

$$c(x,t) \sim t^\theta \phi(x/t^{-1}), \tag{3.33}$$

where exponents $\theta = 1 + p$ is typically fixed by the conservation law and $\phi(\xi)$ is known as the scaling function [17, 20].

To obtain a solution for $c(x,t)$ we now substitute Eq. (3.33) in Eq. (3.22) and find the following equation for the scaling function

$$\xi \frac{d\phi(\xi)}{d\xi} + \left(\xi + (1+p)\right)\phi(\xi) = (1+p)\int_\xi^\infty \phi(\eta)d\eta. \tag{3.34}$$

We thus see that it reduces a partial integro-differential equation for the distribution function $c(x,t)$ into an ordinary integro-differential equation for the scaling function $\phi(\xi)$ and the later equation is much simpler to solve rather than solving the former. To simplify it further we differentiate Eq. (3.34) with respect to ξ and obtain

$$\xi \frac{d^2\phi(\xi)}{d\xi^2} + \left(\xi + (2+p)\right)\frac{d\phi(\xi)}{d\xi} + (2+p)\phi(\xi) = 0, \tag{3.35}$$

which we can re-write as

$$(-\xi)\frac{d^2\phi(\xi)}{d(-\xi)^2} + \left((2+p) - (-\xi)\right)\frac{d\phi(\xi)}{d(-\xi)} - (2+p)\phi(\xi) = 0, \tag{3.36}$$

This is exactly Kumar's confluent differential equation and we write its solution as

$$\phi(\xi) = \,_1F_1(2+p; 2+p; -\xi), \tag{3.37}$$

and its asymptotic solution is

$$\phi(\xi) = \, e^{-\xi}. \tag{3.38}$$

where $_1F_1(a; b; z)$ is known as Kumer's function [4] Using it in Eq. (3.56) we can write the solution for $c(x,t)$ in the long-time limit

$$c(x,t) \sim t^{1+p} e^{-xt}. \tag{3.39}$$

An interesting point about finding solutions using the Buckingham Pi-theorem is that specification of initial condition is not required at any stage. It implies that the solution is universal in the sense that it is independent of initial condition.

3.6 Numerical simulation

We know that we performed the numerical simulation based on the rules depicted in the algorithm (i)–(vi). First, we address the question of how to define time. The time scale in the simulations is defined as the number of attempts to divide an interval regardless of whether the attempt is successful or not. We can verify it by checking whether our numerical data can satisfy Eq. (3.25) or not. In Fig. (3.3) we plot $\log(N(t))$ versus $\log(t)$ for different p. In each case we find a straight line with slope $d_f = p$ which agrees with Eq. (3.25). It means that the way we defined time t is the same as described by Eq. (3.25).

Next, we verify the conservation law. To verify it we label all the existing intervals at a given time, say t_i, in a given realization as $x_1, x_2, ..., x_N$. We then measure the sum of the d_fth power of their sizes at each different time t_i in a single realization which is essentially the same as the d_fth moment of $c(x,t)$, i.e., $M_{d_f}(t)$. Plotting the resulting data in Fig. (3.4) as a function of time t we find it a constant. One interesting thing is that the exact numerical value of this quantity in each different realization is different. However, the average of this ensemble of values is also a conserved quantity although the corresponding numerical value is always equal to one, as shown in Fig. (3.4) by deep black colored filled circles, provided the ensemble size is large enough. Furthermore, to verify Eq. (3.29) we measure mean interval size δ by summing the sizes of all the intervals in a given time and dividing the resulting sum by the corresponding interval number.

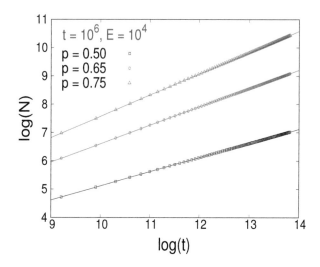

Figure 3.3: (a) Plots of $\log(N)$ versus $\log(t)$ for different p where N is the zeroth moment $M_0(t)$. In each case we find a straight line with slope equal to p as it should be according to Eq. (3.25).

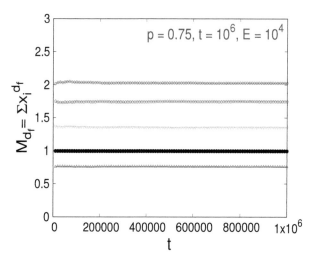

Figure 3.4: The d_fth moment of all the remaining intervals $x_1^{d_f} + x_2^{d_f} + \ldots + x_j^{d_f}$ remain constant in independent realization. However, the numerical value at which it is constant may be different as shown by colored lines. On the other hand, their ensemble averaged value is always equal to one as shown by black colored lines.

We then plot $\log(N)$ versus $\log(\delta)$ in Fig. (3.5) and find an excellent straight line with slope equal to p as predicted by Eq. (3.29).

To test the analytic solution for $c(x,t)$ of the rate equation, given by Eq. (3.22), we perform a simulation for a a given fixed time and then extract the sizes of all intervals. The distribution of interval sizes can then be constructed. We bin the data to find the number of intervals within a suitable class Δ which is then normalized by the width of the class Δ itself so that $c(x,t)\Delta x$ gives the number of intervals which falls within the range x and $x + \Delta x$. We then plot a histogram in Fig. (3.6) where the number of intervals that fall in each class is normalized by the width Δx of the class size. The resulting curves for different times are shown in Fig. (3.7) which represent plots of $c(x,t)$ versus x at four different instant such as at $t_1 = 100k$, $t_2 = 200k$, $t_3 = 300k$ and $t_4 = 400k$ time unit. We then then divide the ordinate by t_i^{1+p} and multiply abscissa by t_i of the $c(x,t)$ vs x data where t_i is the time when the snapshot were taken. The resulting plot, which is equivalent to plots of $t^{-(1+p)}c(x,t)$ vs xt, is shown in Fig. (3.7(a)) which clearly shows that all the distinct plots of Fig. (3.6) now collapsed superbly into one universal curve. Plotting the same curve in the $\log$ $-$ linear scale, as shown in Fig. (3.7(b)), gives a straight line suggesting the solution for the scaling function is exponential as we found analytically.

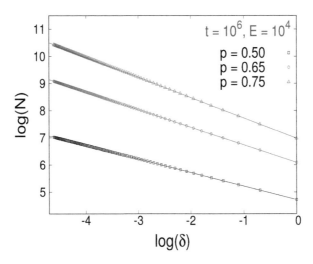

Figure 3.5: We plot $\log(N)$ versus $\log(\delta)$ and find a set of straight lines with slope $d_f = p$. It provides numerical verification of the theoretical relation $N \sim \delta^{-d_f}$ for the stochastic DCS.

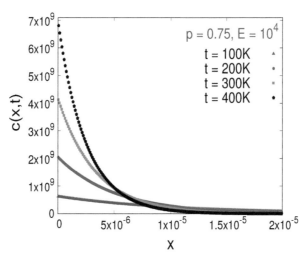

Figure 3.6: The distribution function $c(t, x)$ is drawn as a function of x for three different times. (b) The same set of data are used to plot in the log-linear scale. The straight line clearly proves that $c(x, t)$ for fixed time decays exponentially as predicted by the solution given by Eq. (3.39).

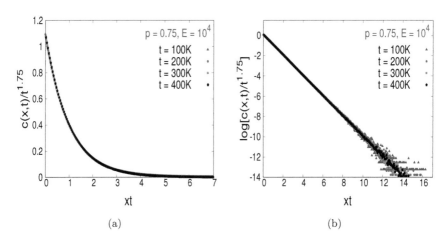

(a)   (b)

Figure 3.7: (a) The distribution function $c(t, x)$ is drawn as a function of x for three different times. (b) The same set of data are used to plot in the log-linear scale. The straight line clearly proves that $c(x, t)$ for fixed time decays exponentially as predicted by the solution given by Eq. (3.56).

Recall that two triangles are called similar even if their dimensional quantities are different though the corresponding dimensionless quantities coincide. For evolving systems like SDCS we can conclude that the system is similar to itself at different times and hence we say that it exhibits self-similarity. Note that self-similarity is also a kind of symmetry as we find that x is replaced by $\lambda^{-1}x$, time t is replaced by λt in Eq. 3.39) we find that $c(x,t)$ can be brought to itself

$$c(\lambda^{-1}x, \lambda t) = \lambda^{1+d_f} c(x,t), \tag{3.40}$$

which is a kind of symmetry. That is, as the system evolves if we take a few snapshots at any arbitrary time and they can all be obtained from one another by similarity transformation. On the other hand, data-collapse, which proves dynamic scaling, mean self-similarity. Actually, self-similarity and similarity transformation are the same thing. All these suggest that there is a continuous symmetry along the time axis. Emmy Noether showed that whenever there exists a continuous symmetry there must exist a conservation law. This is now a well-known Noether's theorem. Indeed, we find that for every value of p within $0 < p < 1$ and β there exists a unique conservation law. That is, the system is governed by a conservation law for each value of β and p which ultimately can be held responsible for fixing the value of θ and z for a given value of albeit all the major variables of the system changes with time. We can thus conclude that the fractal that emerges through evolution in time must obey the Noether's theorem.

3.7 Stochastic fractal in aggregation with stochastic self-replication

The basic idea of the Cantor set is based on fragmentation and removal of intervals in some way. What if we consider its opposite case such as the aggregation of intervals instead of fragmentation. It has been observed that in this case we have to add intervals in some way not remove them to create fractals [83]. In general the kinetics of the aggregation process can be described by Smoluchowski's equation

$$\frac{\partial c(x,t)}{\partial t} = -c(x,t) \int_0^\infty K(x,y)c(y,t)dy$$
$$+ \ \frac{1}{2} \int_0^x K(y, x-y)c(y,t)c(x-y,t)dy, \tag{3.41}$$

where the aggregation kernel $K(x,y)$ determines the rate at which particles of size x and y combine to form a particle of size $(x+y)$ [1, 2]. Essentially, Eq. (3.41) describes the following kinetic reaction scheme,

$$A_x(t) + A_y(t) \xrightarrow{R} A_{(x+y)}(t+\tau), \tag{3.42}$$

where, $A_x(t)$ represents an aggregate of size x at time t and the reaction rate R is related to the aggregation kernel via

$$R = \int_0^\infty K(x, y)c(y, t)dy. \tag{3.43}$$

The factor $1/2$ in the gain term of Eq. (3.41) implies that at each step two particles combine to form one particle. The Smoluchowski equation has been studied extensively in and around the eighties for a large class of kernels satisfying $K(bx, by) = b^\lambda K(x, y)$, where $b > 0$ and λ is known as the homogeneity index [9].

We can consider the case where each of the time two intervals, say, of size x and y combine to form an aggregate of size $x + y$ an interval of the same size is added to the system with probability p. To describe this process we just need to replace the factor $\frac{1}{2}$ of the gain term of Eq. (3.41) by $\frac{1+p}{2}$. The mechanism that the resulting Smoluchowski equation describes can be best understood by giving an algorithm. The process starts with a system that comprise of a large number of chemically identical Brownian particles and a fixed value for the probability $p \in [0, 1]$ by which particles are self-replicated. The algorithm of the model is as follows:

(i)   Two particles, say of sizes x and y, are picked randomly from the system to mimic a random collision via Brownian motion.

(ii) Add the sizes of the two particles to form one particle of their combined size $(x + y)$ to mimic aggregation.

(iii) Pick a random number $0 < R < 1$. If $R \leq p$ then add another particle of size $(x + y)$ to the system to mimic self-replication.

(iv) The steps (i)–(iii) are repeated *ad infinitum* to mimic the time evolution.

One may also consider that the system has two different kinds of particles: active and passive. As the systems evolve, active particles always remain active and take part in aggregation while the character of the passive particles are altered irreversibly to an active particle with probability p. Once a passive particle turns into an active particle it can take part in further aggregation like other active particles already present in the system on an equal footing and it never turns into a passive particle. This interpretation is very similar to the work of Krapivsky and Ben-Naim [32, 48]. While in their work the character of an active particle is altered, in our work it is the other way around. The two models are different also because here we only consider the dynamics of the active particles, whereas Krapivsky and Ben-Naim studied the dynamics of both the entities since a passive particle in their case exists at the expense of an active particle and therefore a consistency check is required. However, the present model does not require such consistency check.

Note that random collision due to Brownian motion can be ensured if we choose a constant kernel $K(x, y)$, e.g.,

$$K(x, y) = 2, \tag{3.44}$$

for convenience. The generalized Smoluchowski equation then is

$$\frac{\partial c(x, t)}{\partial t} = -2c(x, t) \int_0^\infty dy c(y, t) + (1 + p)$$
$$\times \int_0^x dy c(y, t) c(x - y, t). \tag{3.45}$$

This is the fitting equation to the model described by the algorithm $(i) - (iv)$. Incorporating the definition of the nth moment in Eq. (3.45) we obtain

$$\frac{dM_j(t)}{dt} = \int_0^\infty \int_0^\infty dx dy c(x, t) c(y, t) \tag{3.46}$$
$$\times \left[(1 + p)(x + y)^j - x^j - y^j \right].$$

Setting $p = 0$ we can recover the conservation of mass $(M_1(t) = \text{const.})$ of the classical Smoluchowski equation for constant kernel. It is clearly evident from Eq. (3.46) that the mass of the system for $0 < p < 1$ is no longer a conserved quantity, and it is obvious due to the inherent definition of our model. However, it is not obvious from Eq. (3.46) if the system will still be governed by a conservation law or not.

It is fairly easy to obtain solutions to Eq. (3.46) for the first two moments, namely $M_0(t) \equiv N(t)$ and $M_1(t) \equiv M(t)$. For instance, for the mono-disperse initial condition $c(x, 0) = \delta(x - 1)$ they are

$$N(t) = \frac{1}{(1 + (1 - p)t)}, \tag{3.47}$$

and

$$M(t) = L(0)(1 + (1 - p)t)^{\frac{2p}{1-p}}, \quad 0 \le p < 1, \tag{3.48}$$

respectively. We find it convenient to check how the mean or typical interval size $s(t)$ grows with time t. This is defined as

$$s(t) = \langle x \rangle = \frac{\int_0^\infty dx x c(x, t)}{\int_0^\infty dx c(x, t)} = \frac{M_1(t)}{M_0(t)}. \tag{3.49}$$

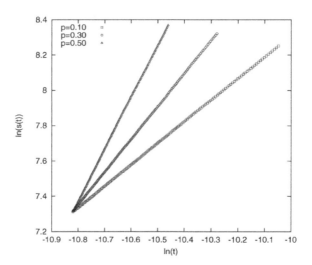

Figure 3.8: We plot $\ln(s(t))$ against $\ln(t)$ for three different values of p starting with mono-disperse initial conditions (we choose $50,000$ particles of unit size). The lines have slopes given by the relation $\frac{1+p}{1-p}$, confirming that $s(t) \sim t^{\frac{1+p}{1-p}}$.

Using Eqs. (3.47) and (3.48) we find that

$$s(t) = \frac{L(0)}{N(0)}(1 + (1-p)N(0)t)^{\left(\frac{1+p}{1-p}\right)}, \quad 0 \le p < 1. \tag{3.50}$$

We thus see that for $0 \le p < 1$ the mean particle size $s(t)$ in the limit $t \to \infty$ grows following a power-law

$$s(t) \sim ((1-p)t)^{\frac{1+p}{(1-p)}}. \tag{3.51}$$

To verify this we plot $\ln(s(t))$ against $\ln(t)$ in Fig. (3.8) using data from numerical simulation for three different values of p with the same mono-disperse initial condition in each case. Appreciating the fact that $t \sim 1/N$ in the long-time limit we obtain three straight lines whose gradients are given by $(\frac{1+p}{(1-p)})$, providing numerical confirmation of the theoretically derived result given by Eq. (3.51).

In fractal analysis, one usually seeks for a power-law relation between the number N needed to cover the object under investigation and a suitable yard-stick size. Let us choose $s(t)$ is the yard-stick size and hence expressing $N(t)$ in terms of $s(t)$ we find

$$N(s(t) \sim s^{-d_f}, \tag{3.52}$$

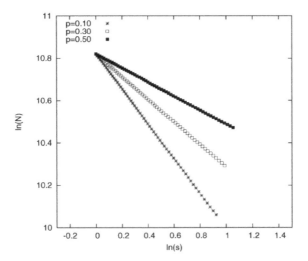

Figure 3.9: Plots of $\ln(N(s))$ against $\ln(s)$ are drawn for three different values of p for the same initial conditions. The lines have slopes equal to $-(\frac{1-p}{1+p})$ as predicted by theory. In each case simulation was performed till $30,000$ aggregation events while the process started with initially $N(0) = 50,000$ particles of unit size.

with $d_f = (1-p)/(1+p)$. Notice that the exponent d_f is a non-integer $\forall\, p$ where $0 < p < 1$ and its value is less than the dimension of the embedding space and hence it is the fractal dimension of the resulting system. To verify our analytical result, we have drawn $\ln(N)$ versus $\ln(s)$ in Fig. (3.9) from the numerical data collected for a fixed initial condition but varying only the p value. On the other hand, in Fig. (3.10) we have drawn the same plots for a fixed p value but varying only initial conditions (monodisperse and polydisperse). Both figures show an excellent power-law fit as predicted by Eq. (3.52) with an exponent exactly equal to d_f regardless of the choice we make for the initial size distribution of particles in the system.

We already know from the Cantor set that the d_fth moment is a conserved quantity. To check this in the present context we label each particle of the system at a given time t by the index $i = 1, 2, 3, ..., N$ where $N = M_0(t)$. Then we construct the d_fth moment at time t given by $\sum_i x_i^{d_f}$ which is equivalent to its theoretical counterpart $\int_0^\infty x^{d_f} c(x, t) dx$ in the continuum limit. Using data from numerical simulation we have shown in Fig. (3.11) that the sum of the qth power of the sizes of all the existing particles in the system remain conserved regardless of time t if we choose $q = \frac{1-p}{1+p}$. Conserved quantities have always attracted physicists as they usually point to some underlying symmetry in the theory

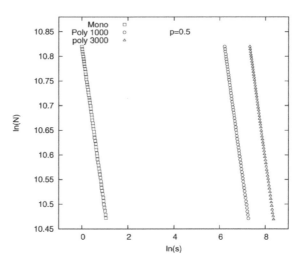

Figure 3.10: The parallel lines resulting from plots of $\ln(N(s))$ against $\ln(s)$ for mono-disperse and poly-disperse initial conditions confirming that $N(s) \sim s^{-\left(\frac{1-p}{1+p}\right)}$ is independent of the initial conditions. In each case simulation started with initially 50,000 particles were drawn randomly from the size range between 1 and 1000 for poly 1000, between 1 and 3000 for poly 3000 and for monodisperse initial condition all the particles were chosen to be of unit size.

or model in which they manifest. Therefore, it is worth pursuing an understanding of the non-trivial value $\frac{1-p}{1+p}$ for $p > 0$ as it leads to the conserved quantity $M_{\left(\frac{1-p}{1+p}\right)}(t)$ in the scaling regime. Such a non-trivial conserved quantity has also been reported in one of our recent works on condensation-driven aggregation and indicates that it is closely related to the fractal dimension. It will be interesting if we find similar close connections between fractal dimension and the non-trivial conserved quantity.

We shall now apply the Buckingham Pi theorem to obtain the scaling solution as it will show us that the resulting fractal is also self-similar. Note that according to Eq. (3.45) the governed parameter c depends on three parameters x, t and p. However, the knowledge about the growth law for the mean particle size implies that one of the parameters, say x, can be expressed in terms of t and p since according to Eq. (3.51) the quantity $((1-p)t)^{\frac{1+p}{(1-p)}}$ bear the dimension of particle size. Note though that p itself does not have dimension, yet we are keeping it as we find it convenient for our future discussion. If we consider $(1-p)t$ as an independent parameter then the distribution function $c(x,t)$ too

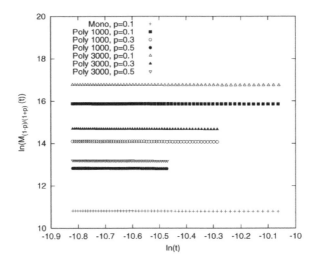

Figure 3.11: $\ln(M_{(\frac{1-p}{1+p})}(t))$ is plotted against $\ln(t)$ for various values of p and various different initial conditions. The horizontal straight lines indicate that $M_{(\frac{1-p}{1+p})}(t)$ is constant in the scaling regime. In all cases initially 50,000 particles were drawn randomly from the size range between 1 and n where $n = 1000, 3000$ and denoted as poly n.

can be expressed in terms of $(1-p)t$ alone, and using the power-law monomial nature of the dimension of physical quantity we can write $c(x,t) \sim ((1-p)t)^\theta$. We therefore can define a dimensionless governing parameter

$$\xi = \frac{x}{((1-p)t)^z},\qquad(3.53)$$

where $z = \frac{1+p}{(1-p)}$ and a dimensionless governed parameter

$$\Pi = \frac{c(x,t)}{((1-p)t)^\theta}.\qquad(3.54)$$

The numerical value of the right hand side of the above two equations remains the same even if the time t is changed by some factor μ for example since the left hand side are dimensionless. It means that the two parameters x and t must combine to form a dimensionless quantity $\xi = x/t^z$ such that the dimensionless governed parameter Π can only depends on ξ. In other words, we can write

$$\frac{c(x,t)}{((1-p)t)^\theta} = f(x/t^z),\qquad(3.55)$$

which leads to the following dynamic scaling form

$$c(x,t) \sim ((1-p)t)^\theta f(x/((1-p)t)^z), \qquad (3.56)$$

where the exponents θ and z are fixed by the dimensional relations $[t^\theta] = [c]$ and $[t^z] = [x]$ respectively and $f(\xi)$ is known as the scaling function.

We now use the scaling form given by Eq. (3.56) into Eq. (3.45) and find that the scaling function $\phi(\xi)$ satisfies

$$
t^{-(\theta+z+1)} = \frac{(1-p)^{2\theta+z}}{F(p,\xi)} \Big[-2\mu_0 f(\xi)
$$
$$
+ (1+p) \int_0^\xi f(\eta)f(\xi-\eta)d\eta \Big], \qquad (3.57)
$$

where

$$F(p,\xi) = \Big[\theta(1-p)^\theta f(\xi) - z(1-p)^\theta \xi \frac{df(\xi)}{d\xi}\Big], \qquad (3.58)$$

and

$$\mu_0 = \int_0^\infty d\xi f(\xi), \qquad (3.59)$$

is the zeroth moment of the scaling function. The right hand side of Eq. (3.57) is dimensionless and hence the dimensional consistency requires $\theta + z + 1 = 0$ or

$$\theta = -\frac{2}{1-p}. \qquad (3.60)$$

The equation for the scaling function $f(\xi)$ which we have to solve for this θ value is

$$(1+p)\Big[\xi\frac{df(\xi)}{d\xi} + \int_0^\xi f(\eta)f(\xi-\eta)d\eta\Big] = 2f(\xi)(\mu_0-1). \qquad (3.61)$$

Integrating it over ξ from 0 to ∞ immediately gives $\mu_0 = 1$ and hence the equation that we have to solve to find the scaling function $f(x)$ is

$$\xi\frac{df(\xi)}{d\xi} = -\int_0^\xi f(\eta)f(\xi-\eta)d\eta. \qquad (3.62)$$

To solve Eq. (3.62) we apply the Laplace transform $G(k)$ of $f(\xi)$ in Eq. (3.62) and find that $G(k)$ satisfies

$$\frac{d}{dk}\Big(kG(k)\Big) = G^2(k). \qquad (3.63)$$

It can be easily solved after linearizing it by making the transformation of the form $G(k) = 1/u(k)$ and integrating straightaway gives

$$G(k) = \frac{1}{1+k}. \tag{3.64}$$

Using it in the definition of the inverse Laplace transform we find the required solution

$$f(\xi) = e^{-\xi}, \tag{3.65}$$

and hence according to Eq. (3.56) the scaling solution for the distribution function is

$$c(x,t) \sim ((1-p)t)^{-\frac{2}{(1-p)}} e^{-x/((1-p)t)^{\frac{1+p}{(1-p)}}}. \tag{3.66}$$

The advantage of using the scaling theory is that one does not need to specify the initial condition revealing the fact that the solution is true for any initial condition.

The question is: How do we verify Eq. (3.66) using the data extracted from the numerical simulation? First, we need to appreciate the fact that each step of the algorithm does not correspond to a one time unit since time $t \sim 1/((1-p)N)$ in the long-time limit as predicted by Eq. (3.47). Secondly, we collect data for a fixed time t and appreciate the fact that $c_t(x)$ is the histogram where the height represents the number of particles within a given range, say of width Δx, is normalized by the width itself so that area under curve gives the number of particles present in the system at time t regardless of their size. This is exactly what is shown in Figs. (3.12) and (3.13) while the Fig. (3.13) is shown in the log-linear scale to show that $c_t(x)$ for fixed time decays exponentially. Now, the solution given by Eq. (3.66) implies that distinct data points of $c(x,t)$ as a function of x at various different times can be made to collapse on a single master curve if we plot $t^{\frac{2}{(1-p)}} c(x,t)$ vs $xt^{-\frac{1+p}{(1-p)}}$ instead. Note that multiplying time t by a constant multiplying factor $(1-p)$ has no impact in the resulting plot. Indeed, we find that the same data points of all the three distinct curves of Figs. (3.12) and (3.13) merge superbly onto a single universal curve, which is essentially the scaling function $f(\xi)$. It is clear from Fig. (3.14) that the scaling function $f(\xi)$ decays exponentially and once again this is in perfect agreement with our analytical solution given by Eq. (3.65). Besides aggregation with stochastic self-replication, particles may also grow in size by condensation, deposition or by accretion. For instance, aggregation in vapor phase or in damp environment particles or droplets may continuously grow by condensation. It has been shown that the condensation-driven aggregation problem can be best described as fractal [66, 68].

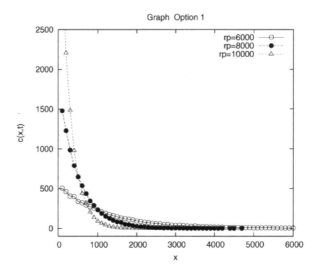

Figure 3.12: Plot of distribution function $c(x,t)$ as a function of x is shown at three different times using data obtained by numerical simulation. Essentially, it is a plot of a histogram where the number of particles in each class size is normalized by the width Δx of the interval size.

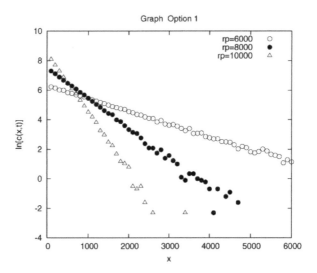

Figure 3.13: Log-linear plot of the same data as in Fig. (3.12) showing the exponential decay of the particle size distribution function $c_t(x)$ with particle size x at fixed time as seen analytically.

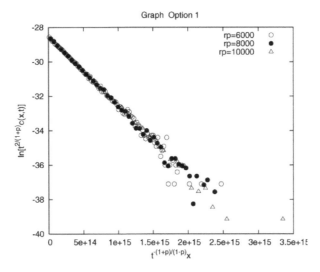

Figure 3.14: The three distinct curves of Figs. (1) and (2) for three different system sizes are well collapsed onto a single universal curve when $c(x,t)$ is measured in units of $t^{-\frac{2}{(1-p)}}$ and x measured in units of $t^{\frac{1+p}{(1-p)}}$. Such data-collapse implies that the process evolves with time preserving its self-similar character. We have chosen semi-log scale to demonstrate that the scaling function decays exponentially $f(\xi) \sim e^{-\xi}$ as predicted by the theory.

3.8   Discussion and summary

We have investigated several fractals of seemingly disparate nature. Despite their differences we have found two common things. First, they are self-similar either spatially or temporally. Second, in each case there is at least one conservation law. Now self-similarity being also a kind of symmetry suggests an intimate connection between which is reminiscent of the Noether's first theorem. We argue that self-similarity cannot exist without a quantity that remains conserved. Think of the stochastic fractal that evolves through time yet each snapshot taken at different times are similar. Here similarity means numerical values of the dimensional quantities of each snapshot may differ but the numerical values of the corresponding dimensionless quantities must remain the same. The conserved quantity is the quantity which preserves the self-similar property.

Chapter 4

Multifractality

4.1 Introduction

Fractals only tell us about an object in which a certain physical quantity is distributed evenly wherever it is found in the embedding space. Therefore, the fractal dimension of the object will still be the same even if the content is diluted to change the density of the content and then distributed throughout the space exactly in the same way as before revealing that it tells us nothing about the distribution of the content. There are situations in which physical quantities are unevenly distributed over the space. That is, the distribution may not be homogeneous and instead heterogeneous in the sense that the density of the content over the occupied regions only are different. Many precious elements such as Gold for instance are found in high concentrations at only a few places, in moderate to lower concentration at many places and in very low concentrations (where the cost of extraction is higher than the price of gold extracted) almost everywhere. In such cases, the fractal *vis-a-vis* the fractal dimension tells us nothing about distribution. We therefore have to invoke the idea of multifractality which tells us about the uneven or heterogeneous distribution of the content on a geometric support. The dimension of the support may itself be a fractal or not.

Perhaps an example can provide better understanding than long description. For this we consider the case of the generalized Sierpinsky carpet which is constructed as follows. Let us consider that the initiator is a square of unit area and unit mass. That is, a unit amount of mass is distributed in square of unit area uniformly. The generator then divides the initiator into four equal smaller squares and the content of the upper left square is pasted onto the lower left square. This process involves cutting and pasting by applying the generator on the available squares over and over again. In the end we have two things: (i) We have the support without content, (ii) We have unevenly distributed mass on the geometric support which is the Sierpinsky carpet. We shall show rigorously in this chapter that the distribution of mass on the support is multifractal. Note that we talk about fractals if all the occupied squares of equal size at any stage have the same amount of mass. It is in this sense we say that the mass is evenly distributed. However, in the case of a cut and paste model the mass is not distributed evenly rather it is distributed unevenly. In this section we develop multifractal formalism in the context of both deterministic as well as a random or stochastic case to be more precise.

So, a system is a candidate for multifractal analysis if a certain physical quantity or variables fluctuate wildly in space or on a support. Consider the case of a stochastic cantor set. In the previous chapter we have seen that the nth moment $M_n(t)$ of the interval size distribution function $c(x, t)$ of stochastic fractal has in general the following form

$$\int_0^\infty x^n c(x, t) dx = M_n(t) \sim t^{-(n-D)z}. \tag{4.1}$$

We therefore can calculate

$$\langle x^n \rangle = \frac{\int_0^\infty x^n c(x, t) dx}{\int_0^\infty c(x, t) dx} = \frac{M_n(t)}{M_0(t)}, \tag{4.2}$$

and

$$\langle x \rangle = \frac{\int_0^\infty x c(x, t) dx}{\int_0^\infty c(x, t) dx} = \frac{M_1(t)}{M_0(t)}. \tag{4.3}$$

Using Eq. (4.1) in Eq. (4.2) and Eq. (4.3) we find

$$\langle x^n \rangle \sim \langle x \rangle^n. \tag{4.4}$$

Thus, there exists a single length scale called the mean interval size $< x >$ which can characterize all the moments of the distribution function. This is true because the exponent $\frac{n - D(\beta)}{3\beta + 2}$ of $M_n(t)$ is linear in n which implies that there exists a constant gap between the exponents of consecutive n value [32]. Such linear dependence of the exponent of the nth moment is taken as the lit-mass test for simple scaling.

There are cases where one finds

$$\langle x^n \rangle \neq \langle x \rangle^n, \tag{4.5}$$

which is typical to the systems where fluctuations in the distribution of the properties of interest is wild in nature. Such systems are typical candidates which can be checked whether it is multifractal or not. One of the important ingredients used to obtain the significance of the spectrum of dimension requires us to use Legendre transformation. We there find it worthwhile to discuss the Legendre transformation in brief.

4.2   The Legendre transformation

To understand the significance of $f(\alpha)$ we need to use the Legendre transform of $\tau(q)$. Before we apply this it is worth it to first understand the principle of Legendre transformation [73]. For this we consider a function

$$y = y(x_1, x_2, ..., x_n). \tag{4.6}$$

Legendre transformation can then be described as a method whereby the derivatives

$$m_k = \left(\frac{\partial y}{\partial x_k} \right), \tag{4.7}$$

can be considered as independent variables without sacrificing any of the mathematical contents of the Eq. (4.6). For the sake of simplicity, let us first consider a function of single variable, e.g.,

$$y = y(x). \tag{4.8}$$

Now,

$$m = \left(\frac{\partial y}{\partial x} \right), \tag{4.9}$$

is the slope of this curve $y(x)$ at the point (x, y). In order to treat m as an independent variable in place of x, one may be tempted to eliminate x between the Eq. (4.8) and Eq. (4.9), thereby getting y as a function of m. However, this is not correct since in doing so one would sacrifice some of the mathematical content of Eq. (4.8) in the sense knowing y as a function of m we will not be able to reconstruct the function $y = y(x)$ as emphasized in Fig. (1).

As an illustrative example, consider the function $y = \frac{1}{2}e^{2x}$ for which $m = (dy/dx) = e^{2x}$ and hence, $y(m) = m/2$. Let us try to reconstruct $y(x)$ from the relation $y(m) = m/2$. Since $m = (dy/dx) = 2y$, we get $y(x) = Ce^{2x}$ where the integration constant C remains unspecified. Clearly, the inadequacy of this procedure arises from the fact that the relation $y = y(m)$ involves dy/dx and the integration required to get $y(x)$ from this first order differential equation yields $y(x)$ up to an unspecified constant of integration.

It is straightforward to realize even from this simple example that a curve (a sequence of points) $y = y(x)$ can also be represented uniquely by the relation $c = c(m)$ that describes the tangent lines. In other words, a knowledge of the intercepts c of the tangent lines as a function of the slope m enables one to construct the family of tangent lines and thence the curve as shown in the Fig. (3). Now the question is: How to calculate $c(m)$ if we are given $y(x)$? The appropriate mathematical tool is known as Legendre transformation. Consider a tangent line that passes through the point x, y and has a slope m and intercept c on the $Y-$ axis. Then,

$$c = y - mx, \tag{4.10}$$

can be taken as the analytical definition of c. In fact, c is the Legendre transform y.

We note that there is a simple and general method for systematically obtaining all the state functions or the thermodynamic potentials describing a system with state variables associated with exact differential forms which is done by using the method of Legendre transformation. Let us first discuss the Legendre transformation in general of a function Y which depends on the variables $x_1, x_2, ..., x_i$, etc. i.e.,

$$Y = Y(x_1, x_2, ..., x_i). \tag{4.11}$$

The partial derivatives of Y with respect to its variables are

$$a_i = \left(\frac{\partial Y}{\partial x_i}\right)_{x_j, j \neq i}. \tag{4.12}$$

If now x_1 is replaced by $a_1 = \left(\partial Y/\partial x_1\right)_{x_j, j \neq 1}$ as independent variable we can then define a new function as

$$\Psi = Y - a_1 x_1, \tag{4.13}$$

and in general

$$\Psi = Y - \sum_i a_i x_i. \tag{4.14}$$

The new transformed state function Ψ is now function of a_i. Such a transformation from the original variables x_i to a new set of variables is called the Legendre transformation.

Once we know the fundamental principle behind the Legendre transformation we can follow the simple working procedure described below without having to go through all the details.

(i) We first have to find the differential dY of the function Y whose Legendre transformation we intend to find. Say it is

$$Y = adx. \tag{4.15}$$

(ii) We then re-write it as

$$dY = adx + xda - xda. \tag{4.16}$$

(iii) Hence we can write

$$d(Y - ax) = -xda. \tag{4.17}$$

(iv) The quantity in the parenthesis of the left hand side of the above equation can be defined as a new quantity

$$\Psi \equiv Y - ax, \tag{4.18}$$

which according to the right hand side of Eq. (4.91) reveals that it is a function of a which is in fact the slope of the original function Y. In the following section we will in fact use this procedure.

4.3   Theory of multifractality

In general, one can describe a set S of points describing an object which can be divided into boxes labeled by an index i such that the ith box will have N_i points of the total N points of the set. These points are sample points describing the content of the underlying measure. Let us use the 'mass' or probability $\mu_i = N_i/N$ in the i-th cell to construct weighted d-measure which we write

$$M_d(q, \delta) = \sum_{i=1}^{N} \mu_i^q \delta^d = Z(q, \delta)\delta^d \tag{4.19}$$

The mass exponent $\tau(q)$ for the set depends on the moment

$$Z(q, \delta) = \sum_i \mu_i^q, \tag{4.20}$$

of order q chosen. It is also known as the partition function as it behaves in a similar manner when one tries to find an analogy of multifractal formalism with thermodynamic phase transition. Like $N(\delta)$ of bare d-measure the partition function (if we may call) $Z(q, \delta)$ also often exhibits power-law distribution

$$Z(q, \delta) \sim \delta^{-\tau(q)}, \tag{4.21}$$

where the exponent $\tau(q)$ is called the mass exponent not the fractal dimension if it is non-linear in q [23]. The mass exponent is more revealing than the simple fractal dimension as we shall soon find out. It can equivalently be defined as

$$\tau(q) = -\lim_{\delta \to 0} \frac{\ln Z(q, \delta)}{\ln \delta}. \tag{4.22}$$

Using Eq. (4.21) in Eq. (4.20) we can write

$$M_d(q, \delta) \sim d^{d-\tau(q)} \xrightarrow{\delta \to 0} \begin{cases} 0 & d > \tau(q) \\ \infty & d < \tau(q) \end{cases}. \tag{4.23}$$

This weighted d-measure has a critical mass exponent $d = \tau(q)$ for which the measure neither vanishes nor diverges as $\delta \to 0$ analogous to the definition of Hausdorff-Besicovitch dimension for the bare d-measure. The weighted measure is characterized by a whole sequence of exponents $\tau(q)$ that controls how the moments of the probabilities $\{\mu_i\}$ scale with δ.

4.3.0.1 Properties of the mass exponent $\tau(q)$

■ We first note that if we choose $q = 0$ then we have $\mu_i^{q=0} = 1$ and hence we find $Z(q = 0, \delta) = N(\delta)$. This number $Z(0, \delta) = N(\delta)$ is simply the number of boxes needed to cover the support and therefore

$$\tau(0) = D, \tag{4.24}$$

is the Hausdorff-Basicovitch dimension of the support and tells nothing about the content in it.

■ On the other hand, the probabilities are normalized: $\sum \mu_i = 1$, and it follows from Eq. (4.22) that

$$\tau(1) = 0. \tag{4.25}$$

Choosing large values of q, say 10 or 100, in Eq. (4.20) favours contributions from cells with relatively high values of μ_i since $\mu_i^q \gg \mu_j^q$, with $\mu_i > \mu_j$ if $q \gg 1$. conversely, $q \ll -1$ favours the cells with relatively low values of the measure μ_i on the cell. These limits are best discussed by considering the derivative $d\tau(q)/dq$ given by

$$\frac{d\tau(q)}{dq} = -\lim_{\delta \to 0} \frac{\sum_i \mu_i^q \ln \mu}{(\sum_i \mu_i^q) \ln \delta}. \tag{4.26}$$

Let $\mu_{\min}$ be the minimum value of μ_i in the sum. Then we find

$$\frac{d\tau(q)}{dq}\bigg|_{q\to-\infty} = -\lim_{\delta\to 0}\frac{(\sum'_i \mu^q_{\min})\ln\mu_{\min}}{(\sum'_i \mu^q_{\min})\ln\delta}, \tag{4.27}$$

where the prime on the sum indicates that only cells with $\mu_i = \mu_{\min}$ contribute. The expression may be written as

$$\frac{d\tau(q)}{dq}\bigg|_{q\to-\infty} = -\lim_{\delta\to 0}\frac{\ln\mu_{\min}}{\ln\delta} = -\alpha_{\max}. \tag{4.28}$$

A similar argument in the limit $q \longrightarrow \infty$ leads to the conclusion that the minimum value of α is given by

$$\frac{d\tau(q)}{dq}\bigg|_{q\to+\infty} = -\lim_{\delta\to 0}\frac{\ln\mu_{\max}}{\ln\delta} = -\alpha_{\min}, \tag{4.29}$$

where $\mu_{\max}$ is the largest value of μ_i, which leads to the smallest value of α. Eqs. (4.28) and (4.29) implies that

$$\mu_{\min} \sim \delta^\alpha_{max}, \tag{4.30}$$

and

$$\mu_{\max} \sim \delta^\alpha_{\min}. \tag{4.31}$$

We can combine the above two relations and write a general equation

$$\mu \sim \delta^\alpha. \tag{4.32}$$

Later we shall show that it is indeed the case and that

$$\alpha(q) = -d\tau/dq, \tag{4.33}$$

is true in general [16, 22].

For $q = 1$ we find that $\frac{d\tau}{dq}$ has an interesting value:

$$\frac{d\tau(q)}{dq}\bigg|_{q=1} = -\lim_{\delta\to 0}\frac{\sum_i \mu_i \ln\mu_i}{\ln\delta} = \lim_{\delta\to 0}\frac{S(\delta)}{\ln\delta}, \tag{4.34}$$

where $S(\delta)$ is the information entropy of the partition of the measure which we may write as

$$S(\delta) = -\sum_i \mu_i \ln\mu_i \sim -\alpha(1)\ln\delta. \tag{4.35}$$

The exponent $\alpha(1) = -(d\tau/dq)|_{q=1} = d_S$ is also the fractal dimension of the set onto which the measure concentrates and describes the scaling with the box size δ of the entropy of the measure. Note that the partition entropy $S(\delta)$ at resolution δ is given in terms of the entropy S of the measure by $S(\delta) = -S\ln\delta$.

4.3.1 *Legendre transformation of $\tau_s(q)$: $f(\alpha)$ spectrum*

Often we find that the Legendre transform of a function is more useful than the function itself. For instance, the Hamiltonian $H = T + V$ which is the Legendre transform of Lagrangian $L = T - V$, the Helmoltz free energy F which is the Legendre transform of the internal energy E, etc. are more useful than their respective function. In the present case we have a function

$$\tau = \tau(q), \tag{4.36}$$

which is non-linear in q and hence its slope depends on q. Legendre transformation of $\tau(q)$ means a method whereby its derivative

$$\frac{d\tau}{dq} = -\alpha, \tag{4.37}$$

can be considered an independent variable instead of q itself [23]. It is straightforward to realize that the curve given by Eq. (4.36) can also be represented uniquely by the relation $f = f(\alpha)$ that describes a family of tangent lines as a function of slope α without sacrificing any of the mathematical content of the parent equation.

Note that the slope of τ vs q should be different at each value of q. From the curve we can draw tangent for a given value of q and write down the equation for the resulting straight line. In general, if we denote α as the slope and the intercept f (which should depend on the slope itself) then the equation for the straight line is

$$\tau = -\alpha q + f(\alpha). \tag{4.38}$$

The function $f(\alpha)$ is in fact the Legendre transform of the function $\tau(q)$. Below, we give a more familiar systematic processing procedure to obtain the Legendre transform of the mass exponent $\tau(q)$. From Eq. (4.37) we can write

$$d\tau = -\alpha dq, \tag{4.39}$$

which we can re-write as

$$d\tau = -\alpha dq - qd\alpha + qd\alpha, \tag{4.40}$$

and hence

$$d(\tau + \alpha q) = qd\alpha. \tag{4.41}$$

The above equation immediately reveals that we can define a new function

$$f(\alpha) = \tau + \alpha q, \tag{4.42}$$

where the new function f is function of α.

4.3.1.1 Physical significance of α and $f(\alpha)$

The sequence of mass exponents $\tau(q)$ is related to the $f(\alpha)$ spectrum via Legendre transformation. Recall the multifractal partition function given by equation

$$Z(q,\delta) = \sum_i \mu_i^q, \qquad (4.43)$$

where cells are indexed by labelling them as $i = 1, 2, ...,$etc. On the other hand, we have seen that the probability μ scales as δ^α. So, the distribution of the content can be subdivided into subsets characterized by α instead of i. We can think of cells which scales sharing the same α. We can now obtain the number $Z(\alpha,\delta)$ needed to cover such subsets of cells. Since the exponent α is continuum variable we further consider that $\rho(\alpha)d\alpha$ is the number of subsets from S_α to $S_{\alpha+d\alpha}$. Upon transition from discrete into continuum variable we write

$$Z(\alpha,\delta)d\alpha = \rho(\alpha)d\alpha\delta^{-f(\alpha)}, \qquad (4.44)$$

and hence the so called partition function can be re-written as

$$Z(\alpha,\delta) = \int \rho(\alpha)d\alpha\delta^{-f(\alpha)}\delta^{\alpha q} = \int \rho(\alpha)d\alpha\delta^{q\alpha-f(\alpha)}. \qquad (4.45)$$

The integral in Eq. (4.45) will be dominated by the value of α which makes $q\alpha - f(\alpha)$ smallest, provided that $\rho(\alpha)$ is nonzero. Thus, we replace α by $\alpha(q)$, which is defined by the extremal condition

$$\frac{d}{d\alpha}[q\alpha - f(\alpha)]\Big|_{\alpha=\alpha(q)} = 0. \qquad (4.46)$$

We also have

$$\frac{d^2}{d\alpha^2}[q\alpha - f(\alpha)]\Big|_{\alpha=\alpha(q)} > 0, \qquad (4.47)$$

so that $f'(\alpha(q)) = q$, $f''(\alpha(q)) < 0$. From Eq. (4.32) it follows that we then have $q = 0$, and we conclude from Eq. (4.13) that $f_{\max} = D$, since $\tau(0) = D$, where D is the fractal dimension of the support of the measure.

4.4 Multifractal formalism in fractal

We already know that the d_fth moment of the interval size at any stage during the construction process is equal to one. It means we can consider each interval as a cell containing a mass equal to the

d_fth power of the size of the interval. If the cells are labelled as $i = 1, 2, \ldots,$, etc. then we can define

$$\mu_i = x_i^{d_f}. \tag{4.48}$$

The corresponding partition function therefore is

$$Z_q = \sum_i \mu_i^q, \tag{4.49}$$

for discrete system and for the continuum system it is

$$Z_q = \int_0^\infty \mu_i^q c(x, t) dx. \tag{4.50}$$

Let us first consider the dyadic Cantor set [85]. In the nth step of its construction there are $(1+p)^n$ number of intervals each of size $x_i = 2^{-n}$ and hence the mean interval size is also $\delta = 2^{-n}$. Thus the partition function is

$$Z_q = \sum_i \mu_i^q = N(2^{-n})^{qd_f} = \delta^{-(1-q)d_f}, \tag{4.51}$$

so that

$$\tau(q) = (1 - q)d_f. \tag{4.52}$$

Its Legendre transformation gives

$$f(\alpha) = d_f, \tag{4.53}$$

and hence it is just a simple fractal.

On the other hand, if we apply the formalism in the stochastic dyadic Cantor set then the partition is

$$Z_q = \int_0^\infty \mu^q c(x, t) dx, \tag{4.54}$$

where $\mu = x^{d_f}$. We can then immediately write

$$Z_q = M_{qd_f}(t). \tag{4.55}$$

We know the solution for the nth moment

$$M_n(t) \sim t^{-(n-d_f)z}. \tag{4.56}$$

Using Eq. (4.56) in Eq. (4.55) gives

$$Z_q \sim t^{-(q-1)d_f z}. \tag{4.57}$$

Expressing it in terms of δ we

$$Z_q \sim \delta^{-(1-q)d_f}. \tag{4.58}$$

Once again the same result as it is for the deterministic dyadic Cantor set [85].

4.4.1 *Deterministic multifractal*

In this section we further modify the construction process of the dyadic Cantor set. Here the support is the DCS with fractal dimension $d_f = \log(1+p)/\log 2$ on which the total mass of the initiator will be distributed heterogeneously. However, before we do that, let us first apply the multifractal formalism to the DCS and show that it results in a unique f value instead of a multifractal $f(\alpha)$ spectrum. Recall that at step one the generator divides the initiator of unit area into two equal pieces and remove one with probability $1-p$. In the next, the generator is applied to each of the remaining intervals to divide them into two equal pieces. If we continue the process then in the nth step we will have $N = (1+p)^n$ intervals of size $s = 2^{-n}$ and we already know that in the large n limit the resulting system emerges as a fractal. However, here we want to apply the multifractal formalism so that we can appreciate the difference between the fractal and multifractal.

In order to apply the multifractal formalism, we first have to know what to choose as a measure and find what fraction of this measure is in the ith cell. To make things simpler let us think of the a case where the generator divides the initiator into two equal parts but removes the right half with probability $1 - p$. That is, after step one there are $1 + p$ intervals of size equal to $1/2$. In such case we assume that each cell is occupied with a content equal to $(1/2)^{d_f}$. In step two, we divide the interval in the left into two equal parts and remove the right half with probability $1 - p$. Similarly, the interval which is there with probability p is also divided in the same fashion. At the end of step two, we shall have $(1+p)^2$ intervals of size $x_i = 1/2^2$. If we now continue the process then at the end of the nth step we shall have $(1+p)^n$ intervals of size $x_i = 2^{-n}$. So, we could describe that the occupation probability of each is $p_i = x_i^{d_f}$ so that

$$\sum_{i=1}^{(1+p)^n} p_i = 1, \tag{4.59}$$

independent of the step number n.

The so called partition function Z_q therefore is

$$Z_q = \sum_{i}^{(1+p)^n} p_i^q. \tag{4.60}$$

Now eliminating n in favour of s where $s = 2^{-n}$ we can write

$$Z_q(s) \sim s^{-(1-q)d_f}. \tag{4.61}$$

The mass exponent therefore is

$$\tau(q) = (1 - q)d_f, \tag{4.62}$$

with $d_f = \ln(1+p)/\ln 2$. The mass exponent satisfies the two conditions (i) $\tau(0) = \frac{\ln 3}{\ln 2}$ and $\tau(1) = 0$. It is worthwhile to mention as a passing note that Hentschel and Procaccia introduced another set of dimensions defined by

$$\tau(q) = (1-q)D_q. \tag{4.63}$$

A system can be described as multifractal only if D_q depends on q. In such case D_0 is just the dimension of the support while D_1 is the information dimension and D_2 is the correlation dimension. Legendre transformation of $\tau(q)$ immediately gives

$$f(\alpha) = \frac{\ln(1+p)}{\ln 2}. \tag{4.64}$$

Thus, instead of getting $f(\alpha)$ spectrum we get a constant value and hence it is not a multifractal since slope of the mass exponent is the same regardless of the value of q. We conclude with the note that the resulting system is the fractal since the fraction of the total measure contains is the same in each cell of same size [97].

Now we slightly change the construction process of the DCS. Instead of throwing away the right half of the interval after each time we apply the generator and paste it onto the left half that always remains with probability $1-p$ [97]. That is, after step one there are $1+p$ intervals of which one in the left has mass $(2-p)/2$ and the other one in the right has mass $p/2$ so that the sum is equal to the mass of the initiator. In

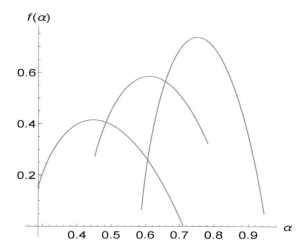

Figure 4.1: Multifractal $f(\alpha)$ spectrum of multifractal dyadic Cantor set for $p = 1/3, 1/2$ and $p = 2/3$. The peak of the respective curves occur at $\log(1+p)/\log(2)$ which is the skeleton on which the measure is distributed.

such a case we could assume that the left cell has the fractional mass equal to $p_1 = (2 - p)/2$ and the right cell has mass equal to $p_2 = 1/2$ with probability p. That is, after step $n = 1$ there are two types of mass $\mu_0 = p_1$ and $\mu_1 = p_2$ such that the total mass $\sum_{k=0}^{1} N_k \mu_k = 1$ is conserved since $N_k = \begin{pmatrix} 1 \\ k \end{pmatrix} p^k$ where $k = 0$, and 1. In step two, the generator is applied to the remaining $1 + p$ intervals and in each we paste the right half onto the left half with probability $(1 - p)$ and hence at the end of step two there are $(1 + p)^2$ intervals with three types of masses $\mu_0 = p^2$ with probability one and two cells have mass $\mu_2 = p_1 p_2$ with probability p and $\mu_3 = p_2^2$ with probability p^2. The total mass of all the cells is $\sum_{k=0}^{2} N_k \mu_k = 1$ where the number of cells of k type is $N_k = \begin{pmatrix} 2 \\ k \end{pmatrix}$ and $k = 0, 1$ and 2. This process of cut and paste is repeated *ad infinitum*. After the nth step we have $(1 + p)^n$ intervals with $n + 1$ different types of mass. We label each type by an integer k ($k = 0, 1., ..., n$) such that the mass of each of the k-type is given by $\mu_k = p_1^{n-k} p_2^k$. The number of the k-type squares N_k is given by $N_k = \begin{pmatrix} n \\ k \end{pmatrix} p^k$ where $k = 0, 1, ..., n$. The distribution of the mass which is clearly not uniform rather heterogeneous. We shall now analyze it below invoking the idea of multifractal formalism.

We can now construct the partition function as

$$Z_q = \sum_{k=0}^{n} N_k \mu_k^q = \sum_{k=0}^{n} \begin{pmatrix} n \\ k \end{pmatrix} \left(\left(\frac{2-p}{2} \right)^q \right)^{n-k} \left(\frac{p}{2^q} \right)^k \tag{4.65}$$

We can re-write it as

$$Z_q = \left(\left(\frac{2-p}{2} \right)^q + \frac{p}{2^q} \right)^n. \tag{4.66}$$

If we now measure the partition function Z_q in terms of $s = 2^{-n}$, then we can eliminate n from Eq. (4.66) in favour of s and find that

$$Z_q = s^{-\tau(q)}, \tag{4.67}$$

where

$$\tau(q) = \frac{\log \left[\left(\frac{2-p}{2} \right)^q + \frac{p}{2^q} \right]}{\log[2]}. \tag{4.68}$$

One can easily check that two essential properties of the mass exponent $\tau(q)$ are obeyed. For instance, $\tau(1) = 0$ and $\tau(0) = \frac{\ln(1+p)}{\ln 2}$ is the dimension of the support. How to obtain the $f(\alpha)$ curve?

To find the fractal subset, we use the usual Legendre transformation of $\tau(q)$. This is a method whereby the slope of $\tau(q)$ is considered an independent variable instead q itself. To obtain the Legendre transformation we write

$$d\tau = \frac{d\tau(q)}{dq} dq, \tag{4.69}$$

where $\frac{d\tau(q)}{dq}$ is slope of the curve τ versus q curve and define it as

$$\alpha(q) = -\frac{d\tau(q)}{dq}. \tag{4.70}$$

We can now write Eq. (4.69) as follows

$$d(\tau + \alpha q) = q d\alpha. \tag{4.71}$$

The above equation immediately reveals that we can define the new function

$$f(\alpha) = \tau(q) + \alpha q, \tag{4.72}$$

where the new function f is function of α since the quantity q is constant on right hand side of Eq. (4.71) while α is variable [97]. In this way q is replaced by α as independent variable. We can thus obtain the following expression for the multifractal spectrum

$$f(\alpha) = q\alpha + \frac{\log\left[\left(\frac{2-p}{2}\right)^q + \frac{p}{2^q}\right]}{\log[2]}, \tag{4.73}$$

and the Hölder exponent is

$$\alpha(q) = \frac{1}{\ln 2} \frac{\frac{p}{2^q} \ln 2 - \left(\frac{2-p}{2}\right)^q \ln\left[\frac{2-p}{2}\right]}{\left(\frac{2-p}{2}\right)^q + \frac{p}{2^q}}, \tag{4.74}$$

The $f(\alpha)$ vs α spectrum is shown in Fig. (4.3) which clearly shows that the pick value of the $f(\alpha)$ is the dimension of the skeleton $\log(1 + p)/\log 2$ and it is concave in shape. Here the skeleton is the dyadic Cantor set but the distribution of the content in this skeleton is multifractal.

4.5   Cut and paste model on Sierpinski carpet

In this section we will first discuss an example of multifractal which is strictly self-similar in character. It is constructed by a recursive process for which exact analytical calculations can be made. Here the support is the Sierpinsky carpet with $b = 2$ which we have already discussed in section 4.3.3. However, let us apply the multifractal formalism to the Sierpinsky carpet and show that it results in a unique f value instead of $f(\alpha)$ spectrum. Recall that at step one the generator divides the initiator of unit area into $b^2 = 4$ equal pieces and removes one which we assume always to be the top left of the four new squares. In the next, the generator is applied to each of the remaining three blocks to divide them into four equal pieces. If we continue the process then in the nth step we will have $N = 3^n$ blocks of size $\delta = 2^{-n}$ and we already know that in the large n limit the resulting system emerges as a fractal of dimension $d_f = \ln 3/\ln 2$. However, here we want to apply the multifractal formalism so that we can appreciate the difference between fractal and multifractal.

In order to apply the multifractal formalism we first have to know what to choose as a measure and find what fraction of this measure is in the ith cell. To make things simpler let us think of the case where the generator divides the initiator into four equal parts but remove nothing. In such case we could assume that each block is occupied with a content equal to its area and the sum of all the areas would be equal to one as the initiator is chosen to be a square of unit area. So, we could describe the area of the respective block as the occupation probability $p_i = x_i^2$ of each block where x_i^2 is the area of the ith block and hence $\sum_i p_i = \sum_{i=1}^{4^n} = 1$. The exponent 2 in $p_i = x_i^2$ is in fact the dimension of the resulting system which is actually the square lattice. We can generalize the idea. In the context of the Sierpinsky carpet we already know its dimension $d_f = \ln 3/\ln 2$. So, in analogy with the square lattice we find that the d_fth moment of the sides of the remaining squares $\sum_{i=1}^{3^n} x_i^{d_f}$ too is a conserved quantity. We, therefore, can assume that ith block of 3^n remaining block is occupied with a certain content equal to $p_i = x_i^{\ln 3/\ln 2}$. It can be easily shown that indeed $\sum_{i=1}^{3^n} p_i = 1$ since

$$x_1 = x_2 = \ldots = x_{3^n} = 2^{-n} \tag{4.75}$$

and hence

$$\sum_{i=1}^{3^n} x_i^{d_f} = 3^n 2^{-nd_f} = e^{n\ln 3} e^{-nd_f \ln 2} = 1. \tag{4.76}$$

We therefore can once again regard p_i as the occupation probability. The so called partition function Z_q therefore is

$$Z_q = \sum_{i}^{3^n} p_i^q. \tag{4.77}$$

Now eliminating n in favour of δ where $\delta = 2^{-n}$ we can write

$$Z_q(\delta) \sim \delta^{-(1-q)d_f}. \tag{4.78}$$

The mass exponent therefore is

$$\tau(q) = (1 - q)d_f, \tag{4.79}$$

where $d_f = \ln 3 / \ln 2$ which is the fractal dimension of the Sierpiksy carpet. The mass exponent satisfies the two conditions (i) $\tau(0) = \frac{\ln 3}{\ln 2}$ and $\tau(1) = 0$. It is worthwhile to mention as a passing note that Hentschel and Procaccia introduced another set of dimensions defined by

$$\tau(q) = (1 - q)D_q. \tag{4.80}$$

A system can be described as multifractal only if D_q depends on q. In such case D_0 is just the dimension of the support while D_1 is the information dimension and D_2 is the correlation dimension. Legendre transformation of $\tau(q)$ immediately gives

$$f(\alpha) = \frac{\ln 3}{\ln 2}. \tag{4.81}$$

So, it is not a multifractal since the slope of the mass exponent is the same regardless of the value of q. We conclude with the note that since the fraction of the total measure contained is the same in each cell of same size. In the cut and paste model we will take the Sierpinsky carpet as the support where blocks are occupied by a fraction of the total measure which can be different even if the cells are of same size.

The construction of the cut and paste model can be described by a curdling process where the initiator is a square of unit area and of unit mass distributed on the square uniformly [44]. Then the generator of the Sierpinsky carpet is modified as follows. In step one then the generator divides the initiator into four equal squares and upper left square is removed and paste its mass equal to 1/4 on the lower left square. The mass of the lower left square is therefore equal to 1/2 and each of the remaining two squares has mass equal to 1/2. In step two the generator is applied to the remaining three squares and in each case mass from their upper left squares are removed and paste their mass on the respective lower left squares. This process of cut and paste is repeated *ad infinitum*. The distribution of the mass which is clearly not even rather heterogeneous which we have analyzed invoking the idea of multifractal formalism discussed in the previous section.

The construction process may equivalently be described as follows. At the first step ($n = 1$), divide the square into four equal squares; redistribute the total mass of the original square into three smaller

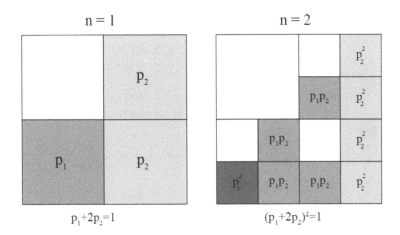

Figure 4.2: The first two steps are shown to illustrate the cut and paste model on the Sierpinski carpet.

squares of which the two on the right have a fractional mass $p_2 = 1/4$ and one on the bottom left has mass $p_1 = 1/2$ such that $p_1 + 2p_2 = 1$. That is, after step $n = 1$ there are two types of mass $\mu_0 = p_1$ and $\mu_1 = p_2$. Repeat this process recursively for each of the smaller squares. The case of $n = 2$ is shown in Fig (2). So, in step $n = 2$ the generator when applied to the lower left square of μ_0 will now have two types of mass $\mu_0 = p_1^2$ and $\mu_1 = p_1 p_2$. Similarly, when the generator is applied to the upper right squares then the bottom left of the four new smaller squares will have mass $\mu_1 = p_2 p_1$ and the two squares on the right will have mass $\mu_2 = p_2^2$. In the same way, when the generator will be applied on the third square will again have two types of mass as obtained for the upper right square. Thus at step $n = 2$ we find three types of mass. It can be shown that at the n-th step, the linear size of each square is $\delta = 2^{-n}$, and there are $(n+1)$ different types of squares when classified according to their masses. Each type can be designated by an integer k $(k = 0.1., ..., n)$ such that the mass of each of the k-type squares is given by $\mu_k = p_1^{n-k} p_2^k$. The number of the k-type squares N_k is given by $N_k = 2^k \binom{n}{k}$.

These results can easily be seen by observing that the different masses in the n-th step can be obtained in a multiplicative process. At the n-th step, the possible masses in the squares are given by all the terms in the expression of $(p_1 + p_2)^n$, which is equivalent to

$$\sum_{k=0}^{n} \binom{n}{k} p_1^{n-k}(2p_2)^k = \sum_{k=0}^{n} 2^k \binom{n}{k} p_1^{n-k}p_2^k. \tag{4.82}$$

How to obtain the mass exponent? The fraction of masses in the i-th cell μ_i can be taken to be the fractal measure. The sequence of mass exponents $\tau(q)$ is defined by

$$N(q,\delta) = \sum_{i} \mu_i^q \sim \delta^{-\tau(q)}, \tag{4.83}$$

as $\delta \longrightarrow 0$, where δ is the size of each cell. Specialized to our case here, one obtains

$$\begin{aligned}
N(q,\delta) &= \sum_{k=0}^{n} N_k \mu_k^q \\
&= \sum_{k=0}^{n} 2^k \binom{n}{k} (p_1^{n-k}(p_2^k)^k)^q \tag{4.84} \\
&= \sum_{k=0}^{n} \binom{n}{k} (p_1^q)^{n-k}(2p_2^q)^k) \\
&= (p_1^q + 2p_2^q)^n.
\end{aligned}$$

Eliminating n in favour of δ using $\delta = 2^{-n}$ and using Eqs (4.82) and (4.83), we find that

$$Z_q(\delta) \sim \delta^{-\tau(q)}, \tag{4.85}$$

where

$$\tau(q) == [\ln(p_1^q + 2p_2^q)]/\ln 2. \tag{4.86}$$

One can easily check that two essential properties of the mass exponent $\tau(q)$ are obeyed. For instance, $\tau(1) = 0$ and $\tau(0) = \frac{\ln 3}{\ln 2}$ is the dimension of the support. How to obtain the $f(\alpha)$ curve? Using Legendre transformation and after a few steps of algebra gives the following expression for the multifractal spectrum

$$f(\alpha) = q\alpha + [\ln(p_1^q + 2p_2^q)]/\ln 2, \tag{4.87}$$

and the Hölder exponent is

$$\alpha(q) = -\frac{1}{\ln 2} \frac{p_1^q \ln p_1 + 2p_2^q \ln p_2}{p_1^q + 2p_2^q}. \tag{4.88}$$

It only requires us to see the difference between the cut and paste model and the Sierpinski carpet in order to understand the origin of multifractality.

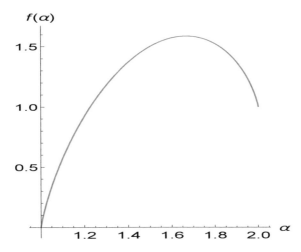

Figure 4.3: Multifractal $f(\alpha)$ spectrum of multifractal dyadic Cantor set for $p = 1/3, 1/2$ and $p = 2/3$. The peak of the respective curves occur at $\log(1+p)/\log(2)$ which is the skeleton on which the measure is distributed.

4.6 Stochastic multifractal

4.7 Weighted planar stochastic lattice model

Perhaps, the square lattice is the simplest example of the cellular structure where every cell has the same size and the same coordination number. Its construction starts with an initiator, say a square of unit area, and a generator that divides it into four equal parts. In the next step and steps thereafter the generator is applied to all the available blocks which eventually generates a square lattice. In this section, we intend to address the following questions. Firstly, we ask what if the generator is applied to only one of the available blocks at each step by picking it preferentially with respect to the areas? Secondly, we ask what if we use a modified generator that divides the initiator randomly into four blocks instead of four equal parts and apply it to only one of the available blocks at each step, which are again picked preferentially with respect to their respective areas? Our primary focus will be on the later case which results in the tiling of the initiator into increasingly smaller mutually exclusive rectangular blocks. We term the resulting structure as *weighted planar stochastic lattice* (WPSL) since the spatial randomness is incorporated by the modified generator and also the time is incorporated in it by the sequential application of the modified generator [75, 80, 92]. The definition of the model may appear too simple but the results it offers, as we shall see soon, are far from simple. To illus-

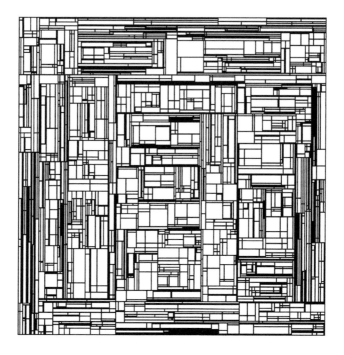

Figure 4.4: A snapshot of the weighted planar stochastic lattice containing 30001 blocks.

trate the type of systems expected, we show a snapshot of the resulting weighted planar stochastic lattice taken during the evolution (see Fig. (4.4)). We intend to investigate its topological and geometrical properties in an attempt to find some order in this seemingly disordered lattice.

4.8   Algorithm of the weighted planar stochastic lattice (WPSL)

Perhaps an exact algorithm can provide a better description of the model rather than the mere definition. In step one, the generator divides the initiator, say a square of unit area, randomly into four smaller blocks. The four newly created blocks are then labelled by their respective areas a_1, a_2, a_3 and a_4 in a clockwise fashion starting from the upper left block (see Fig. 2). In step two and thereafter only one block is picked at each step with the probability equal to their respective area and then it is divided randomly into four blocks. In general, the jth step of the algorithm can be described as follows.

(i) Subdivide the interval $[0, 1]$ into $(3j - 2)$ sub-intervals of size $[0, a_1]$, $[a_1, a_1 + a_2]$, ..., $[\sum_{i=1}^{3j-3} a_i, 1]$ each of which represents the blocks labelled by their areas $a_1, a_2, ..., a_{(3j-2)}$ respectively.

(ii) Generate a random number R from the interval $[0, 1]$ and find which of the $(3i - 2)$ sub-interval contains this R. The corresponding block it represents, say the pth block of area a_p, is picked.

(iii) Calculate the length x_p and the width y_p of this block and keep note of the coordinate of the lower-left corner of the pth block, say it is (x_{low}, y_{low}).

(iv) Generate two random numbers x_R and y_R from $[0, x_p]$ and $[0, y_p]$ respectively and hence the point $(x_R + x_{low}, y_R + y_{low})$ mimicking a random nucleation of a seed in the block p.

(v) Draw two perpendicular lines through the point $(x_R + x_{low}, y_R + y_{low})$ parallel to the sides of the pth block mimicking orthogonal cracks parallel to the sides of the blocks which stops growing upon touching existing cracks and divide it into four smaller blocks. The label a_p is now redundant and hence it can be reused.

(vi) Label the four newly created blocks according to their areas a_p, $a_{(3j-1)}$, a_{3j} and $a_{(3j+1)}$ respectively in a clockwise fashion starting from the upper left corner.

(vii) Increase time by one unit and repeat the steps (i)–(vii) *ad infinitum*.

In general, the distribution function $C(x, y; t)$ describing the blocks of the lattice by their length x and width y evolves according to the following kinetic equation [32]

$$\frac{\partial C(x, y; t)}{\partial t} = -C(x, y; t) \int_0^x \int_0^y dx_1 dy_1 F(x_1, x - x_1, y_1, y - y_1) + \quad (4.89)$$

$$4 \int_x^\infty \int_y^\infty C(x_1, y_1; t) F(x, x_1 - x, y, y_1 - y) dx_1 dy_1,$$

where kernel $F(x_1, x_2, y_1, y_2)$ determines the rules and the rate at which the block of sides $(x_1 + x_2)$ and $(y_1 + y_2)$ is divided into four smaller blocks whose sides are the arguments of the kernel [31, 32, 33]. The first term on the right hand side of equation (4.89) represents the loss of blocks of sides x and y due to nucleation of seeds of crack on one such block from which mutually perpendicular cracks are grown to divide it into four smaller blocks. Similarly, the second term on the right hand side represents the gain of blocks of sides x and y due to nucleation of seeds of crack on a block of sides x_1 and y_1 ensuring that one of the four new blocks have sides x and y. Let us now consider the case where the generator divides the initiator randomly into four smaller rectangles and apply it to only one of the available squares thereafter by picking preferentially with respect to their areas. It effectively describes the random sequential nucleation of seeds with uniform probability on the initiator. Within the rate equation approach this can be ensured if one chooses the following kernel

$$F(x_1, x_2, y_1, y_2) = 1 \tag{4.90}$$

Substituting it into equation (4.89) we obtain

$$\frac{\partial C(x, y; t)}{\partial t} = -xyC(x, y; t) + 4 \int_x^\infty \int_y^\infty C(x_1, y_1; t) dx_1 dy_1. \tag{4.91}$$

The coefficient xy of $C(x, y; t)$ of the loss term implies that seeds of cracks are nucleated on the blocks preferentially with respect to their areas which is consistent with the definition of our model.

Incorporating the 2-tuple Mellin transform given by

$$M(m, n; t) = \int_0^\infty \int_0^\infty x^{m-1} y^{n-1} C(x, y; t) dx dy, \tag{4.92}$$

in equation (4.91) we get

$$\frac{dM(m, n; t)}{dt} = \left(\frac{4}{mn} - 1\right) M(m+1, n+1; t). \tag{4.93}$$

Iterating equation (4.93) to get all the derivatives of $M(m, n; t)$ and then substituting them into the Taylor series expansion of $M(m, n; t)$ about $t = 0$ one can immediately write its solution as

$$M(m, n; t) = {}_2F_2\left(a_+, a_-; m, n; -t\right), \tag{4.94}$$

where $M(m, n; t) = M(n, m; t)$ for symmetry reason and

$$a_\pm = \frac{m+n}{2} \pm \left[\left(\frac{m-n}{2}\right)^2 + 4\right]^{\frac{1}{2}}, \tag{4.95}$$

where ${}_2F_2\left(a, b; c, d; z\right)$ is the generalized hypergeometric function [4].

One can see that (i) $M(1,1;t) = 1 + 3t$ is the total number of blocks $N(t)$ and (ii) $M(2,2;t) = 1$ is the sum of areas of all the blocks which is obviously a conserved quantity [35, 41]. Both properties are again consistent with the definition of the WPSL depicted in the algorithm. The behaviour of $M(m,n;t)$ in the long time limit is

$$M(m,n;t) \sim t^{-a_-}. \tag{4.96}$$

Thus, in addition to the conservation of total area the system is also governed by infinitely many non-trivial conservation laws as it implies

$$M(n,4/n;t) \sim \text{constant} \quad \forall n. \tag{4.97}$$

We used numerical simulation to verify equation (4.97) or its discrete counterpart $\sum_i^N x_i^{n-1} y_i^{4/n-1}$ if we label all the available blocks as $i = 1, 2, ..., N$. We found that the analytical solution is in perfect agreement with the numerical simulation which we performed based on the algorithm for the WPSL model (see Fig. 3).

4.9   Geometric properties of WPSL

We now find it interesting to focus on the distribution function $c(x,t) = \int_0^\infty f(x,y,t)dy$ that describes the concentration of rectangles which have

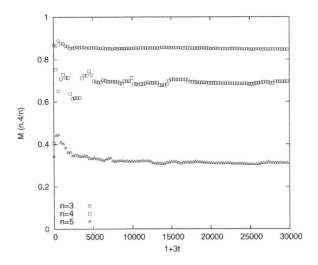

Figure 4.5: The plots of $\sum_i^N x_i^{n-1} y_i^{4/n-1}$ vs N for $n = 3, 4, 5$ are drawn using data collected from one realization.

length x at time t regardless of the size of their widths y. Then the qth moment of $n(x,t)$ is defined as

$$M_q(t) = \int_0^\infty x^q n(x,t)dx. \tag{4.98}$$

$$M_q(t) \sim t^{\{\sqrt{q^2+16}-(q+2)\}/2}. \tag{4.99}$$

Note that for symmetry reasons it does not matter whether we consider the qth moments of $n(x,t)$ or that of $n(y,t)$ since we have $M(q+1,1;t) = M(1,q+1;t)$. According to equation (4.99) the quantity $M_3(t)$ and hence $\sum_i^N x_i^3$ or $\sum_i^N y_i^3$ is a conserved quantity. Although $M_3(t)$ remains a constant against time in every independent realization the exact numerical value is found to be different in every different realization (see Fig. 4). Like in the dyadic Cantor set their ensemble average value is, however, equal to one. We also find that the qth moment of $c(x,t)$ and the qth power of the first moment are not equal, i.e.,

$$<x^q> = \frac{\int_0^\infty x^q n(x,t)dx}{\int_0^\infty n(x,t)dx} \neq <x>^q = \left(\frac{\int_0^\infty xn(x,t)dx}{\int_0^\infty n(x,t)dx}\right)^q. \tag{4.100}$$

It suggests that a single length scale cannot characterize all the moments of the distribution function $n(x,t)$.

4.9.0.1 Multifractal analysis to stochastic Sierpinski carpet

Of all the conservation laws we find that $M_3(t) = \sum_i x_i^3$ is a special one since we can use it as a multifractal measure consisting of members p_i, the fraction of the total measure $p_i = x_i^3/\sum_i x_i^3$, distributed on the geometric support WPSL. That is, we assume that the ith block is occupied with the cubic power of its own length x_i. The corresponding "partition function" of multifractal formalism then is

$$Z_q(t) = \sum_i p_i^q \sim M_{3q}(t). \tag{4.101}$$

Its solution can immediately be obtained from equation (4.99) to give

$$Z_q(t) \sim t^{\{\sqrt{9q^2+16}-(3q+2)\}/2}. \tag{4.102}$$

Using the square root of the mean area $\delta(t) = \sqrt{M(2,2;t)/M(1,1;t)} \sim t^{-1/2}$ as the yard-stick to express the partition function Z_q gives the

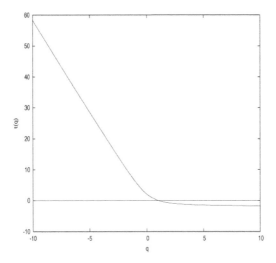

Figure 4.6: The plots of $\tau(q)$ vs q to show its slope as a function of q varies.

weighted number of squares $N(q, \delta)$ needed to cover the measure which we find decays following power-law

$$N(q, \delta) \sim \delta^{-\tau(q)}, \qquad (4.103)$$

where the mass exponent

$$\tau(q) = \sqrt{9q^2 + 16} - (3q + 2). \qquad (4.104)$$

The non-linear nature of $\tau(q)$, see Fig. 5 for instance, suggests that the gap exponent

$$\Delta = \tau(q) - \tau(q - 1) \qquad (4.105)$$

is different for every q value. It implies that we require an infinite hierarchy of exponents to specify how the moments of the probabilities $\{p\}$s scales with δ. Now if we choose $q = 0$ then it gives an estimate of the number of squares $N(0, \delta) = N(\delta)$ of sides δ we need to cover the support on which the members of the population is distributed. We find that $N(\delta)$ scales as

$$N(q = 0, \delta) \sim \delta^{-\tau(0)}, \qquad (4.106)$$

where $\tau(0) = 2$ is the Hausdorff-Besicovitch dimension of the support. On the other hand, if we choose $q = 1$ we have $Z_1 = \sum_i p_i = const.$ and hence we must have $\tau(1) = 0$. This is indeed the case according to equation (4.104). Therefore, $\tau(0) = 2$ (the dimension of the support) and $\tau(1) = 0$ (required by the normalization condition) are often considered as the first self-consistency check for the multifractal analysis.

4.9.0.2 Legendre transformation of the mass exponent $\tau_s(q)$: The $f(\alpha)$ spectrum

We now perform the Legendre transformation of the mass exponent $\tau(q)$ by using the Lipschitz-Hölder exponent, as given in equation (4.37), as an independent variable to obtain the new function

$$f(\alpha) = q\alpha + \tau(q). \tag{4.107}$$

Replacing $\tau(q)$ in equation (4.103) in favour of $f(\alpha)$ we find that

$$N(q(\alpha), \delta) \sim \lim_{\delta \to 0} \delta^{q\alpha - f(\alpha)}. \tag{4.108}$$

On the other hand, using $p \sim \delta^\alpha$ in the expression for partition function $Z_q = \sum_i p^q$ and replacing the sum by integral while indexing the blocks by a continuous Lipschitz-Hölder exponent α as variable with a weight $\rho(\alpha)$ we obtain

$$N(q(\alpha), \delta) \sim \int \rho(\alpha) d\alpha N(\alpha, \delta) \delta^{q\alpha}, \tag{4.109}$$

where $N(\alpha, \delta)$ is the number of squares of side δ needed to cover the measure indexed by α. Comparing Eqs. (4.108) and (4.109) we find

$$N(\alpha, \delta) \sim \delta^{-f(\alpha)}. \tag{4.110}$$

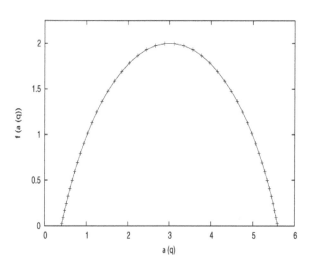

Figure 4.7: The $f(\alpha)$ spectrum.

It implies that we have a spectrum of spatially intertwined fractal dimensions

$$f(\alpha(q)) = \frac{16}{\sqrt{9q^2 + 16}} - 2, \tag{4.111}$$

are needed to characterize the measure. That is, the size disorder of the blocks are multifractal in character since the measure $\{p_\alpha\}$ is related to the size of the blocks. That is, the distribution of $\{p_\alpha\}$ in WPSL can be subdivided into a union of fractal subsets each with fractal dimension $f(\alpha) \leq 2$ in which the measure p_α scales as δ^α. Note that $f(\alpha)$ is always concave in character (see Fig. 6) with a single maximum at $q = 0$ which corresponds to the dimension of the WPSL with empty blocks.

On the other hand, we find that the entropy $S(\delta) = -\sum_i p_i \ln p_i$ associated with the partition of the measure on the support by using the relation $\sum_i p_i^q \sim \delta^{-\tau(q)}$ in the definition of $S(\delta)$. Then a few steps of algebraic manipulation reveals that $S(\delta)$ exhibits scaling

$$S(\delta) = \ln \delta^{-\alpha(1)} \tag{4.112}$$

where the exponent $\alpha_1 = \frac{6}{5}$ obtained from

$$\alpha(q) = - \left. d\tau(q)/dq \right|_q . \tag{4.113}$$

It is interesting to note that $\alpha(1)$ is related to the generalized dimension D_q, is also related to the Rényi entropy $H_q(p) = \frac{1}{q-1} \ln \sum_i p_i^q$ in the information theory, given by

$$D_q = \lim_{\delta \to 0} \left[\frac{1}{q-1} \frac{\ln \sum_i p_i^q}{\ln \delta} \right] = \frac{\tau(q)}{1-q}, \tag{4.114}$$

which is often used in the multifractal formalism as it can also provide insightful interpretation. For instance, $D_0 = \tau(0)$ is the dimension of the support, $D_1 = \alpha_1$ is the Renyi information dimension and D_2 is known as the correlation dimension. Multifractal analysis was initially proposed to treat turbulence but later successfully applied in a wide range of exciting field of research. For instance, it has been recently found that the wild fluctuations of the wave functions at the Anderson and the quantum Hall transition can be best described by multifractality. Recently, though this has got renewed momentum as it has been found that the probability density function at the Anderson and the quantum Hall transition exhibits multifractality since in the vicinity of the transition point fluctuations are wild - a characteristic feature of multifractal behaviour.

4.10 Multifractal formalism in kinetic square lattice

In an attempt to understand the origin of multifractality we now consider the case where the generator divides the initiator into four equal blocks instead of randomly into four blocks. If the generator is applied over and over again thereafter to only one of the available squares by picking preferentially with respect to their areas then it results in the kinetic square lattice (KSL). Within the rate equation approach it can be described by the kernel

$$F(x_1, x_2, y_1, y_2) = (x_1 + x_2)(y_1 + y_2)\delta(x_1 - x_2)\delta(y_1 - y_2). \quad (4.115)$$

and hence the resulting rate equation can be obtained after substituting it in equation (4.89) to give

$$\frac{\partial C(x, y; t)}{\partial t} = -\frac{1}{4}xyC(x, y; t) + 4^2 xyC(2x, 2y; t). \quad (4.116)$$

Incorporating equation (4.92) in equation (4.116) yield

$$\frac{dM(m, n; t)}{dt} = -\left(\frac{1}{4} - \frac{4}{2^{m+n}}\right)M(m + 1, n + 1; t). \quad (4.117)$$

To obtain the solution of this equation in the long-time limit, we assume the following power-law asymptotic behaviour of $M(m, n; t)$ and write

$$M(m, n; t) \sim A(m, n)t^{\theta(m+n)}, \quad (4.118)$$

with $\theta(4) = 0$ since the total area obtained by setting $m = n = 2$ is an obvious conserved quantity. Using it in equation (4.117) yields the following difference equation

$$\theta(m + n + 2) = \theta(m + n) - 1. \quad (4.119)$$

Iterating its subject to the condition that $\theta(4) = 0$ gives

$$M(m, n; t) \sim t^{-\frac{m+n-4}{2}}. \quad (4.120)$$

Apparently it appears that in addition to the conservation of the total area $M(2, 2; t)$, we find that the integrals $M(3, 1; t)$ and $M(1, 3; t))$ are also conserved. Interestingly, all the three integrals $M(2, 2; t)$, $M(3, 1; t)$ and $M(1, 3; t)$ effectively describe the same physical quantity since all the blocks are square in shape and hence

$$\sum_{i=1}^{N} x_i^2 = \sum_{i=1}^{N} y_i^2 = \sum_{i=1}^{N} x_i y_i. \quad (4.121)$$

Therefore, in reality the system obeys only one conservation law - conservation of total area.

We again look into the qth moment of $n(x,t)$ using equation (4.98) and appreciating the fact that $M_q(t)$ equal to $M(q+1,1;t)$ or $M(1,q+1;t)$ we immediately find that

$$M_q(t) \sim t^{-\frac{q-2}{2}}. \tag{4.122}$$

Unlike in the previous case where the exponent of the power-law solution of the $M_q(t)$ is non-linear, here we have an exponent which is linear in q. It immediately implies that in the case of a kinetic square lattice

$$< x^q >=< x >^q, \tag{4.123}$$

and hence a single length-scale is enough to characterize all the moments of $n(x,t)$. That is, the system now exhibits simple scaling instead of multiscaling. Like before let us consider that each block is occupied with a fraction of the measure equal to square of its own length or area $p_i = \sum_i^N x_i^2$ and hence the corresponding partition function is

$$Z_q = \sum_i^N p_i^q = M_{2q}(t). \tag{4.124}$$

Using equation (4.122) we can immediately write its solution

$$Z_q(t) \sim t^{-\frac{2q-2}{2}}. \tag{4.125}$$

Expressing it in terms of the square root of the average area $\delta \sim t^{-1/2}$ gives the weighted number of square $N(q,\delta)$ of side δ needed to cover the measure which has the following power-law solution

$$N(q,\delta) \sim \delta^{-(1-q)2}, \tag{4.126}$$

where mass exponent $\tau(q)$ is

$$\tau(q) = 2 - 2q. \tag{4.127}$$

The Legendre transform of the mass exponent is a constant

$$f(\alpha) = 2, \tag{4.128}$$

and so is the generalized dimension

$$D_q = 2. \tag{4.129}$$

We thus find that if the generator divides the initiator into four equal squares and we apply it thereafter sequentially then the resulting lattice is no longer a multifractal. The reason is that the distribution of the population in the resulting support in this case is uniform. The two models therefore provide a unique opportunity to look for possible origin of multifractality. The two models discussed in this section differs only in the definition of the generator. In the case when the generator divides the initiator randomly into four blocks and we apply it over and over again sequentially then we have multifractality since the underlying mechanism in this case is governed by random multiplicative process. This is not, however, the case if the generator divides the initiator into equal four blocks and applies it over and over again sequentially since the resulting dynamic is governed by deterministic multiplicative process instead.

4.10.1   *Discussions*

In this section, we discuss two main features, (i) finding the explicit time dependent and scaling properties of the particle size distribution function, when particles are characterized by more than one variable and (ii) the connection between the kinetics of the fragmentation process and the occurrence of multifractality to describe the rich pattern formed due to the breakup process. Both these results have important applications including a unique opportunity to search for the origin of multifractality and multiscaling. In reality, fragmenting objects will have both size and shape, i.e., a geometry. Intrigued by the possibility that the geometry of the fragmenting objects may influence the fragmentation process, we have investigated three distinct models of fragmentation. For some simple choices of the fragmentation rule we give exact and explicit solutions to these geometric models.

We find it difficult to find the explicit solution for general homogeneity in the case of two-dimensions. Nevertheless, we find that the solution of the rate equation for the moments is analytically tractable to find the temporal behaviour of the moment in the long time limit. Since the moment keeps the signature of some generic features of the particle size distribution function we confined ourselves to the asymptotic behaviour of the moment for general homogeneity indices which is essential to look out for the occurrence of the shattering transition. We suggest that the existence of an infinite number of hidden conserved quantities clearly indicates the absence of scaling solutions.

The models we discuss in this chapter can also be used to produce stochastic fractals which are reminiscent of the Cantor gasket $(d = 2)$, Cantor cheese $(d = 3)$, etc. We derive exact expressions for fractal dimensions when the initiator is a rectangle and at each time the step generator subdivides into four rectangles and one or more of them are removed randomly. We continue the process *ad infinitum* with the remaining pieces. In this case it appears that one cannot describe the phenomenon by a single fractal dimension-infinitely many are required. Such phenomena are called *multifractal*. Physically, it means that it is possible to partition the resulting system into subsets such that each subset is a fractal with its own characteristic dimension. Typically, multifractal patterns appear in systems that develop far from equilibrium and that do not yield a minimum energy configuration, such as diffusion limited aggregation, or a metal foil grown by electrodeposition.

When the system describes the fragmentation process, the resulting set in the long time limit is an integer number typical for Euclidean shape. Yet, we find that the system shows multifractality and gives a unique measure of support $(D_f = 2)$ on which subsets can be distributed. However, any observable fluctuates strongly from one realization to the other. Although each realization is statistically self-similar in these fluctuations, it means that averaged quantities of any observables can be measured with reasonable accuracy only through the ensemble average. That is, a single experiment over a longer period of time will not give any averaged quantities with good accuracy, instead a large number of independent experiments is required. This is a very important property for real or numerical experiments. But when describing stochastic fractals, one associates pictures of wildly varying probabilities of the measure since at each realization the dimension of the support can be different. This reflects the fact that in the case of a system describing stochastic fractals, the entropy of the system has one more source than in the fragmentation process. This extra source arises due to the competition among the fractal support for different m^* in a given experiment. Recently, fragmentation of a heavy drop falling in a lighter miscible fluid has been performed in the laboratory. During this process, the droplet sizes were found to display multifractal properties which is in agreement with our investigation.

One might wonder if it is possible to say when to expect a random fractal or multifractal. Perhaps, it is quite safe at this stage to say that whenever we hopelessly fail to produce an identical copy under the same initial condition, but each realization has the same generic form to recognize them, we find a fractal object. On the other hand, we expect a system to show multifractality whenever each copy appears with strong fluctuations between different copies and each copy can be partitioned into subsets such that each subset scales with different exponents, yet they can be recognized due to their generic features. Note that when three fragments are removed from the system at each time event $(s = 1)$ the dimension of the measure support $(D_f(s))$ is zero where the measure can be distributed.

Chapter 5

Fractal and Multifractal in Stochastic Time Series

5.1 Introduction

In the previous chapters, we have discussed the concept of self-similarity, fractality and multifractality for both the deterministic and stochastic processes. In each case, a rule (deterministic or stochastic) is considered to generate the process and define its fractal or multifractal behaviour. The deterministic rule is mathematical and the corresponding process is called deterministic [47, 49]. On the other hand, a stochastic rule is completely statistical, i.e; behaviour of the system depends on a probabilistic law and the associated system is known as a stochastic process [50, 53]. However, these kinds of predefined rules do not work well on prediction of real world phenomena (e.g.; human heart oscillation, neuro system, tumor cell growth, cancer growth, environment, socio-economical system, etc). Moreover, it is impossible to define an exact rule of the corresponding phenomena. In such a case, the only way is to predict the process is by time series analysis. A time series is deterministic or stochastic due to the corresponding nature of the process. For the deterministic time series, fractal as well as multifractal analysis can be done by reconstruction of the attractor from the given time series [47, 49]. Reconstruction of the attractor requires a suitable time-delay and proper embedding dimension [14, 47, 49]. On the other hand, such delay embedding method cannot be applied for a stochastic time series (STS) as because both delay and embedding are very large [50]. However, self-similarity can be investigated by measuring a scaling law for the STS. The scaling law can be measured by the fluctuation analysis based on statistical variation in the time series [30, 25, 69, 81]. A single scaling behaviour corresponds to monofractal STS [50]. Perhaps, this single scaling is not always effective to describe the different types of irregularity in the time series. In this context, multifractality is investigated for the STS [50]. In this chapter, we have discussed the method of fluctuation analysis, fractality and multifractality in STS with various numerical examples.

A summarized plan of the discussion is given in Fig. 5.1. In Section 5.2, basic conceptions about scaling law, monofractal and multifractal STS are discussed. As two different fluctuation theories were developed for stationary and non-stationary STS respectively, we, first, discussed the method of verifying the stationarity and non-stationarity of a time series in Section 5.2. The next discussion, given in Section 5.4, is about the fluctuation analysis for monofractal STS. For the stationary monofractal STS, auto-correlation coefficient method (Section 5.4.1), spectral analysis (Section 5.4.2), method of Hurst exponent (Section 5.4.3), fluctuation analysis (Section 5.4.4) can be applied to find the underlying scaling law. On the other hand, non-stationary monofractal STS can be analyzed by wavelet analysis, detrended fluctuation analysis (Section 5.4.5), detrended moving average technique, centred moving average, etc. Perhaps, detrended fluctuation analysis are generally used to characterize the monofractal STS due to some drawbacks and huge computational cost of other methods. Section 5.5 discusses the method of wavelet transform modulus maxima, multifractal detrended fluctuation analysis to characterize a multifractal STS. In the last section, we summarize the chapter and discuss the future scope of all the analyses.

5.2 Concept of scaling law, monofractal and multifractal time series

A collection of random variables $\{X_t\}, t \in T$, where X_t is defined as a map from a probability space (S, P, Ω) to a measurable space $\mathcal{M}$, is known as a stochastic or random process. The index set T is generally considered as time domain. We can get discrete as well as the continuous stochastic process according to $T = \mathbb{N}$ and $\mathbb{R}$ respectively. For example, let us consider a random variable $X : S \to \mathbb{R}$ (S being sample space of

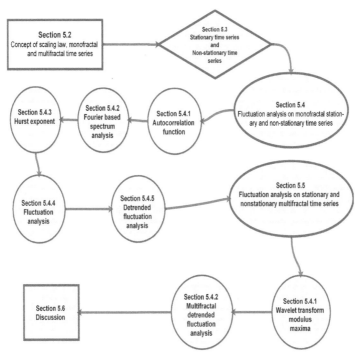

Figure 5.1: A flow chart represents a plan of discussion of the remaining topics. The chart starts from upper left corner (the rectangle) and end at bottom left corner (the square). Each geometrical figure signifies different sections and arrow indicates directions of the progression of the successive sections.

throwing two coins) is considered by

$$X(w_i) = \begin{cases} \text{number of 'head's,} & \text{if } w_i \in S \\ 0, & \text{otherwise.} \end{cases}$$

If S is defined by $\{(H, H) = w_1, (H, T) = w_2, (T, H) = w_3, (T, T) = w_4\}$, then the values of X are given by $X(w_1) = 2, X(w_2) = 1, X(w_3) = 1, X(w_4) = 0$. Then probability P of an event $0 \le X(w_i) \le 1$ denoted by $P(0 \le X \le 1)$ or $P(X \le 1)$ and it's equal to $P(X \le 1) = \frac{1}{2}$.

For any stochastic process, each random variables can make a set of observations at different time $t \in T$. These are known as state spaces of the stochastic process.

Definition 5.1 For a random variable X, a set of observation $\{x(k)\}_{k=1}^{N}$ (sometimes denoted by $\{x\}$) obtained at successive time $t = nk, n \in \mathbb{N}$ is known as stochastic time series (STT) of the corresponding process.

For fixed n, the time series is known as regularly sampled; otherwise we call this irregularly sampled. In this chapter, we dealt with only regularly sampled time series.

Definition 5.2 For a given time series $\{x\}$, a measure μ is said to obey a scaling law with an exponent α if $\mu(cs) = c^\alpha \mu(s)$ hold, where c is a constant and s represents scale parameter.

For example, consider a time series with a measure μ that satisfies the relation $\mu(s) = as^\alpha$. Then, $\mu(cs) = a(cs)^\alpha = c^\alpha \mu(s)$. Thus μ obeys the scaling law with exponent α.

It can be seen that the measure is independent of scaling up to the desired order. So long-term behaviour of the time series can be characterized by a statistical measure that obeys a scaling Laws. Hence, long-term dynamics of a STT can be described by scaling laws of some statistical measures which are valid for a wide range of time or frequencies scales. The parameter α is called self-similarity parameter. According to a single or multiple non-integer scaling exponent, a time series is characterized by monofractal or multifractal process respectively.

Definition 5.3 A stochastic process is said to be a fractal process if the corresponding time series $\{x\}$ obeys the following conditions:
(a) it satisfies scaling law,
(b) the scaling exponent should be non-integer.

Sometimes, it has been observed that single scaling exponent is not sufficient to describe the stochastic process. In that case, time series are needed to be characterized by multiple scaling exponents and the underlying processes are called mutifractal process.

Definition 5.4 A stochastic process is said to be a multifractal process if the corresponding time series $\{x\}$ obeys the following conditions:
(a) it satisfies scaling law with various number of different scaling exponents,
(b) the scaling exponents are non-integer.

In order to find the fractal scaling law, stationarity and non-stationarity are needed to investigate from a given time series. Stationarity, as well as the non-stationarity of a time series, can be verified by the auto-correlation method. The auto-correlation method is basically a correlation of a time series with itself. So, the next section is given for a discussion on the method of verifying stationarity and non-stationarity of a given STS.

5.3 Stationary and non-stationary time series

Stationary time series means its distribution changes nowhere in any time span [24]. For a time series $\{x(k)\}_{k=1}^{N}$ with probability $p(x(k))$, is said to be stationary if and only if $p(x(k)) = p(x(k+\tau))$, where $\tau(\in \mathbb{Z})$ represents translation or lag of the time index k. In a practical situation, equality between $p(x(k))$ and $p(x(k+\tau))$ is not always possible to establish a given time series $\{x(k)\}_{k=1}^{N}$. It is an ideal case that can discuss in the theory of stationary time series. It often call strictly (or strongly) stationarity of a time series. To overcome this limitation, an alternative criteria is proposed which is, in fact, known as weakly stationarity.

Definition 5.5   A time series $\{x(k)\}_{k=1}^{N}$ is said to be weakly stationary if $\langle x \rangle = \langle x(k+\tau) \rangle$ and $\sigma_{x(k)} = \sigma_{x(k+\tau)}$, where

$$\langle x \rangle = \frac{1}{N} \sum_{k=1}^{N} x(k), \quad \sigma_{x(k)} = \sqrt{\frac{1}{N} \sum_{k=1}^{N} \{x(k) - \langle x \rangle\}^2}, \tag{5.1}$$

$\tau(\in \mathbb{Z})$ being the time-delay.

Based on parametric and non-parametric techniques, two methods are generally used to test the stationarity. The parametric method is generally used in the time domain, such as economists, who are making certain assumptions about the nature of financial data [50]. On the other hand, the non-parametric method is mainly developed on the frequency domain, such as engineers, who often consider the system as a black box. In fact, the non-parametric method does not consider any assumption about the system. Moreover, consideration of the normal distribution of the data is not needed in the non-parametric method. Though the non-parametric method is widely applicable from a statistical view, it reveals less powerful results than parametric method. The parametric method is mainly based on auto-covariance or auto-correlation method. We now define auto-covariance and auto-correlation of a STS of length N.

Definition 5.6   Let $\{x(k)\}_{k=1}^{N}$ be a given time series. Then, auto-covariance of $\{x(k)\}_{k=1}^{N}$ with delay τ is denoted by $ACOV(\tau)$ and defined as

$$ACOV(\tau) = \frac{1}{N} \sum_{k=1}^{N-\tau} (x(k) - \langle x \rangle)(x(k+\tau) - \langle x(k+\tau) \rangle), \tag{5.2}$$

where $\tau = 0, 1, 2, 3, \ldots$.

Similarly, we can define $ACOV(-\tau)$ by

$$ACOV(-\tau) = \frac{1}{N} \sum_{k=1-\tau}^{N} (x(k) - \langle x \rangle)(x(k+\tau) - \langle x(k+\tau) \rangle), \qquad (5.3)$$

where $\tau = -1, -2, -3, \ldots$.

Definition 5.7 For a time series $\{x(k)\}_{k=1}^{N}$, auto-correlations $AC(\tau)$ and $AC(-\tau)$ are defined by

$$AC(\tau) = \frac{ACOV(\tau)}{\sigma_{x(k)}\sigma_{x(k+\tau)}}, \qquad \tau = 0, 1, 2, 3, \ldots \qquad (5.4a)$$

$$AC(-\tau) = \frac{ACOV(-\tau)}{\sigma_{x(k)}\sigma_{x(k+\tau)}}, \qquad \tau = -1, -2, -3, \ldots \qquad (5.4b)$$

From the Definition 5.7, it can be verified that a time series will be stationary if and only if $AC(\tau) = AC(-\tau)$. Otherwise, a time series is called non-stationary. A numerical illustration is given in Example 5.1 to show the applicability of auto-correlation measure on verifying the stationarity and non-stationarity of the STS.

Example 5.1 *First consider two different time series x and y, each of length 500, as given in Fig. 5.2. From the Fig. 5.2a, only one variation can be observed in the time series x. It indicates that, the corresponding probability distributions are the same in each time window that covers at least one vibration. So, x is weakly stationary. On the other hand, variable oscillation can be observed in y over different time intervals. In fact, oscillation in y for the time interval $I_1 = (250, 500]$ is faster than the same*

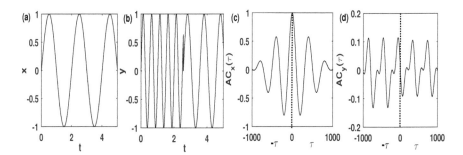

Figure 5.2: (a), (b) represent a stationary time series $\{x\}$ and a non-stationary time series $\{y\}$ respectively. In each case, we have considered same length $L = 500$. (c) and (d) represent respective auto-correlation curves with $\tau \in [-1000, 1000]$. The dotted lines indicates auto-correlations at $\tau = 0$.

in $I_2 = [0, 250]$ *(see Fig. 5.2b). It implies that the respective oscillating patterns in I_1 and I_2 are always differs. As no common oscillation exists there, corresponding probability distribution will change over time and hence non-stationarity indicates in $\{y\}$.*

Self-similarity can be observed in both stationary and non-stationary STS. So both the methods of monofractal and multifractal scaling law can be investigated from the stationary and non-stationary time series respectively. However, fluctuations are needed to be defined separately in each of the cases. The next sections discuss various methods of fluctuation analysis for monofractal STS.

5.4 Fluctuation analysis on monofractal stationary and non-stationary time series

5.4.1 Autocorrelation function

For a given STS $\{x\}$, the autocorrelation function of lag τ was already described in (5.4). Now, if the data is uncorrelated (e.g., a random walk), $AC(\tau) = 0$ for $\tau > 0$. If the data has short term correlation, then $AC(\tau)$ decay exponentially with respect to $\frac{\tau}{t_c}$, where t_c is the characteristic decay time. Moreover, for the long-range correlation $AC(\tau)$ satisfies the condition $AC(\tau) \sim \tau^{-\gamma}$ $(0 < \gamma < 1)$. The exponent γ is known as autocorrelation exponent. In practice, value of γ can be calculated by measuring the gradient of the straight line fitted on $\log AC(\tau)$ vs. $\log \tau$ plot. However, a direct calculation of $AC(\tau)$ on non-stationary time series always reveals incorrect results due to continuous changes in average $\langle x \rangle$. Further, values of $AC(\tau)$ highly fluctuates around zero for large τ which makes it impossible to find out the correct γ [69]. A numerical illustration about the effectiveness of $AC(\tau)$ is given at the end of the Section 5.4.

5.4.2 Fourier based spectrum analysis

For any given continuous STS $\{x\}$, the Fourier spectrum $S(f)$ (f represents frequency of the STS) is defined by

$$S(f) = \|X(f)\|^2, \text{where} \tag{5.5}$$

$X(f)$ is known a Fourier transform of $\{x\}$ and is given by

$$X(f) = \int_{-\infty}^{\infty} x e^{-2\pi t} dt. \tag{5.6}$$

The discrete power spectrum can also be defined by discrete Fourier transform.

For a given stationary time series, scaling behaviour of the time series can be observed by measuring the fluctuation of $S(f)$ over long-range of f. To do this, slope of the straight line is fitted on $\log S(f)$ vs. $\log f$ plot is calculated. The slope is equivalent to β, where β satisfies $S(f) \sim f^{-\beta}$. Further, it has been established that β is related to the autocorrelation exponent γ by $\beta = 1 - \gamma$. So uncorrelated, short-term as well as long-term correlation of an STS can also be characterized by spectrum analysis.

We now discuss efficiency of spectrum analysis in Example 5.2.

Example 5.2 *First consider two different STS-white noise ($\{x\}$) and pink noise ($\{y\}$) shown in Fig. 5.3a and b respectively. From the Fig. 5.3a, it can be observed that fluctuation in the corresponding STS is completely random. On the other hand, less random fluctuation can be seen in Fig. 5.3b. Then, calculate $S(f)$ using (5.5). To compute power β for each STS, we have drawn $\log S(f)$ vs. $\log f$ plots. The corresponding plots are given in Fig. 5.3c and d respectively. By fitting straight lines on the mean trend of the plots, the slope of the respective lines are found to be 0 and 1 respectively.*

Note

In order to apply fluctuation analysis, one of the major tasks is to classify the noisy and random walk nature from the given STS. The method of spectrum analysis (described in Section 5.4.2) can be used to identify these natures. Figure 5.4 shows some STS with $\beta \in [0, 2]$. An STS with $\beta \in [0, 1]$ is known as noisy time series. On the other hand, a random walk can be assured by observing the value of β in $(1, 2]$. Moreover, $\beta = 0$ and 1 indicate white and pink noise respectively.

However, this analysis does not reveal better results, until a logarithmic binning procedure is applied to the double logarithmic plot of $S(f)$.

5.4.3 Hurst exponent

The method of Hurst exponent was proposed by a British hydrologist H.E. Hurst, while he was working in Egypt on the Nile River Dam Project [3]. He observed that both the axis of time and a statistical quantity are not equivalent for an observation or time series. Therefore, rescaling of time is necessary for adjusting the time series. He rescaled the time scale $k \in \mathbb{N}$ by a factor 'a' and the corresponding

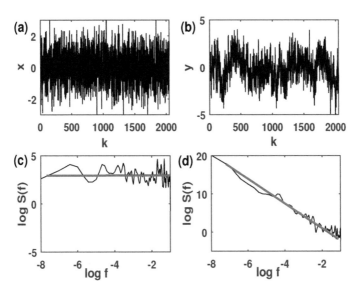

Figure 5.3: (a), (b) represent white noise ($\{x\}$) and pink noise ($\{y\}$) respectively. In each case, length of noise is considered N=2000. (c) and (d) represent corresponding $\log S(f)$ vs. $\log f$ plots. The solid line (pink) indicates fitted straight lines on the $\log S(f)$ vs. $\log f$ plots.

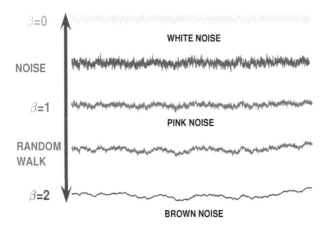

Figure 5.4: Power noise signals $\frac{1}{f^{\beta}}1$ are represents for for $\beta = 0$ (gray), 0.5 (blue), 1 (pink), 1.25 (red), 1.5 (brown).

time series $\{x(k)\}_{k=1}^{N}$ by $a^H x(ak)$ (i.e; $a^H x(ak) \to x(k)$), which actually reflects statistical self-similarity (self-affinity) of the time series. Then he has computed the exponent H based on a statistical quantity applied to the rescaled time series, which is known as the Hurst Exponent (HE). The steps of calculating the HE from a given STS are given as follows:

Step-1: For a given $\{x(k)\}_{k=1}^{N}$, $N_s (N_s = [\frac{N}{s}])$ non-overlapping segments A_ν $(\nu = 1, 2, \ldots, N_s)$ of size s is defined by

$$A_1 = \{x(1), x(2), \ldots, x(s)\}$$
$$A_2 = \{x(s+1), x(s+2), \ldots, x(2s)\}$$
$$\vdots \qquad \vdots$$
$$A_{N_s} = \{x(\overline{N_s - 1}s + 1), x(\overline{N_s - 1}s + 2), \ldots, x(\overline{N_s - 1} + N_s)\}.$$

(5.7)

Step-2: For each segment A_ν $(\nu = 1, 2, \ldots, N_s)$, means $\langle x_\nu \rangle$ and deviations S_ν are calculated by

$$\langle x_\nu \rangle = \frac{1}{s} \sum_{k=1}^{\nu s} x(\overline{\nu - 1}s + k), \quad S_\nu = \sqrt{\frac{1}{s} \sum_{k=1}^{\nu s} \{x(\overline{\nu - 1}s + k) - \langle x_\nu \rangle\}^2}.$$

(5.8)

Step-3: In the next, define two quantities-profile $(X_{\nu j})$ and range $(R_\nu(s))$ by

$$X_{\nu j} = \sum_{k=1}^{j} \{x_{(\nu-1)s+k} - \langle x_\nu \rangle\},$$

(5.9)

$$R_\nu(s) = max X_{\nu j} - min X_{\nu j}, j = (\nu - 1)s + 1, \ldots, \nu s.$$

(5.10)

Step-4: Then, rescale range is obtained by averaging the fluctuation function $F_{RS}(s)$, where

$$F_{RS}(s) = \frac{1}{N_s} \sum_{\nu=1}^{N_s} \frac{R_\nu(s)}{S_\nu(s)}.$$

(5.11)

It can be verified that, $F_{RS}(s)$ obeys the scaling law: $F_{RS}(s) \sim s^H$, where H is known as HE. In practice, H is calculated by fitting a straight line on the $\log F_{RS}(s)$ vs. $\log s$ plot. In fact, slope of the fitted straight line on the $\log F_{RS}(s)$ vs. $\log s$ plot gives the value of H. Example 5.2 illustrates calculations of H for three types of power noise:

Example 5.3 *Power noise is a kind of stochastic time series whose power spectrum $S(f)$ (f being frequency of the time series) obeys the law: $S(f) \sim f^\beta$, where $\beta \in \mathbb{R}^+ \cup \{0\}$. For our purpose, we consider $\beta = 0$, 0.5 and 1. To calculate H, we have investigated $\log s$ vs. $\log R/s(s)$ graphs for the respective aforesaid time series. The corresponding plots are shown in Fig. 5.5 a, b and c respectively. From the figures, it can be investigated that the slope of the fitted straight lines on the corresponding plots are found to be $0.4959, 0.7495$ and 0.8302 respectively. These correspond to the values of H of the respective power noises. It is noted that the values of H for $\beta = 0$ and 0.5 (given in Fig. 5.5a and b) are almost equivalent to its standard theoretical values. For $\beta = 0.5$, the value of H is 0.8302 which is far from its theoretical value (see Fig. 5.5c). However, non-stationarity is one of the major reasons for this incoherence.*

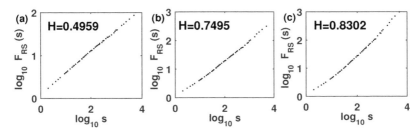

Figure 5.5: (a), (b) and (c) represents $\log_{10} F_{RS}(s)$ vs. $\log_{10} s$ plot for $\beta = 0$, 0.5 and 1 respectively. For each plot, length of the noise are considered 2000. In order to find the HE, straight lines are fitted on the linear region of $\log_{10} F_{RS}(s)$ vs. $\log_{10} s$ plots. On computation, we have chosen scale as 10^s, where $s = [0, 1, 2, 3, 4]$.

Another measure based on random walk theory, known as Fluctuation analysis (FA), was also proposed to find the scaling law fluctuation of the STS. The method of FA is discussed in the following section:

5.4.4 Fluctuation analysis (FA)

In this method, a time series $\{x(k)\}_{k=1}^N$ with zero mean (if not, then mean is needed to make zero) is considered. In order to find fluctuation from the series $\{x(k)\}$, the following steps are to be done:

Step-1: A profile $y(j)$ is constructed by

$$y(j) = \sum_{k=1}^{j} x(k) \quad \text{for} \quad j = 1, 2, \ldots, N. \tag{5.12}$$

The profile $y(j)$ is nothing but a random walker on a linear chain after time step j.

Step-2: Define N_s $(= [\frac{N}{s}])$ segments A_ν $(\nu = 0, 1, 2, \ldots, N_s)$ as given in Eq. (5.7).

Step-3: On each A_ν, square-fluctuation of the profile $y(j)$ is calculated by

$$F^2(\nu, s) = [y(\overline{\nu - 1}s + 1) - y(\nu s)]^2, \quad (\nu = 1, 2, \ldots, N_s) \qquad (5.13)$$

Step-4: Then mean fluctuation of $\{F^2(\nu, s)\}$ $((\nu = 1, 2, \ldots, N_s))$ is calculated by

$$F^2(s) = \frac{1}{N_s} \sum_{\nu=1}^{N_s} F^2(\nu, s). \qquad (5.14)$$

From the Eq. (5.14), it is seen that $F(s)$ is the root-mean-square displacement of the random walker $\{y(j)\}$ at the scale s. For a given time series having long-term correlations, the corresponding fluctuation $F(s)$ increases with some power of s, say s^α. So, we can write $F(s) \sim s^\alpha$.

It can be observed that fluctuation $\{F(s)^{\frac{1}{2}}\}$ follows the scaling law: $\{F(s)^{\frac{1}{2}}\} \sim s^\alpha$. Since $\alpha \approx H$ holds for a monofractal time series, so $2\alpha = 1 + \beta$. In FA, values of α are always taken in $(0, 1)$. At $\alpha = 0, 1$, significant inaccuracy has been observed in FA results. Also, it can be seen that FA is reliable only when s satisfies $s > N/10$.

To overcome all the limitations of FA method, detrended fluctuation analysis (DFA) was proposed by Pend et al. The next section describes the DFA method with numerical illustrations.

5.4.5   *Detrended fluctuation analysis*

Detrended Fluctuation method can characterize the long-term correlation of the non-stationary signal. In DFA, the detrending technique is applied to the linear trend observed in the FA. As detrending applies to the method of FA, it indicates that the DFA also counts the variance of the random walk of the given time series. To calculate detrended fluctuation from a time series $\{x(k)\}_{k=1}^{N}$, we need to perform following steps:

Step-1: A profile $X(j)$ $(j = 1, 2, \ldots, N)$ is constructed from $\{x(k)\}_{k=1}^{N}$ by

$$X(j) = \frac{1}{N} \sum_{k=1}^{j} x(k) - <x>, \text{where} \qquad (5.15)$$

$\langle x \rangle$ denotes the mean of $\{x(k)\}$.

Step-2: Define $N_s(= [N/s])$ non-overlapping segments A_ν of length s same as given in Eq. (5.7).

Step-3: For each segment, a polynomial trend $X_{\nu,s}(j)$ is fitted by least-square method and subtracted from the corresponding original profile. The resultant profile $\tilde{X}_s(j)$ is calculated by $\tilde{X}_s(j) = X(j) - X_{\nu,s}(j)$. [see the Example 5.4]

Step-4: The detrended variance is defined by

$$F^2(s) = \frac{1}{sN_s} \sum_{\nu=1}^{N_s} \sum_{j=(\nu-1)s+1}^{\nu s} [\tilde{X}_s(j)]^2. \qquad (5.16)$$

If the long-term exists in $\{x(k)\}_{k=1}^{N}$, then $F^2(s)$ increases with s according to power law, i.e; $F^2(s) \sim s^\alpha$. The exponent α is called the scaling exponent and it is calculated by measuring the slope of the fitted straight line on $\log F^2(s)$ vs. $\log s$ plot.

Note

Before calculating the values of α, we discuss the detrending technique (described in step-2). To do this, we choose $\frac{1}{f}$-noise and construct a profile $X(j)$ by (5.15). Figure 5.6 a and b show same profile $X(j)$ for $\frac{1}{f}$-noise. Then, we divide the profile on some non overlapping intervals of same length. In Fig. 5.6a, we have considered straight lines for fitting the profile $X(j)$ in each of the intervals. On the other hand, quadratic polynomials are fitted for $X(j)$ in Fig. 5.6b. From both the figures, it can be observed that the best fitting can be obtained in Fig. 5.6b. It indicates, that DFA cannot reveal an appropriate result until the detrending shows minimum errors.

To show the effect of DFA on monofractal analysis, a numerical investigation is done on three power noises. The discussion is given in Example 5.4.

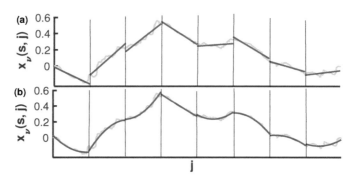

Figure 5.6: (a), (b) represents profile $X(j)(j = 1, 2, \ldots, 8000)$ (green) for a $\frac{1}{f}$-noise of length 8000. In both the figures, dotted (red) lines represents trend $X_\nu(s, j)$. In (a), $X_\nu(s, j)$ represents a straight line for each ν. On the other hand, $X_\nu(s, j)$ represents a quadratic polynomial for each ν in (b). The vertical (black) lines indicates non-overlapping segments.

Example 5.4 *By considering the quadratic trend, we have computed α for each $\beta = 0, 0.5, 1$ using (5.16). Figure 5.7 the respective αs. It can be observed that, values of $\alpha \approx 0.5, 0.75$ and 1 for the aforesaid noises. These values also correspond to the receptive HEs. Moreover, it can be verified that $\alpha = 0.5, > 0.5, < 0.5$ indicates white noise, persistent and anti-persistent behaviour respectively. If $\alpha = 1$ and 1.5, signals look like as $\frac{1}{f}$-noise and Brownian motion. So, unlike the HE, α shows equivalent behaviour. So, this numerical illustration indicates that the DFA is very effective towards finding the scaling behaviour of both stationary and non-stationary STS.*

However, the supremacy of DFA cannot be assured in this stage. To do this, a comparative study between DFA and all the remaining measures for monofractal STS is needed. As DFA is an extension of FA, we only compare the method of DFA with two independent methods autocorrelation function and Fourier based spectrum analysis. For this purpose, we have considered three types of power noise $\frac{1}{f^0}$, $\frac{1}{f^{0.5}}$, and $\frac{1}{f}$ respectively. Figure 5.8a shows the auto-correlation of the respective noise. It can be observed from the figure that, all the fluctuations having a similar trend with respect to the scale s. On the other hand, corresponding trends of the $\log f$ vs. $\log S(f)$ are all equivalent for the three respective power noises (see Fig. 5.8b). So, the behaviour of the noise cannot always be possible to distinguish by the method of auto-correlation and Power spectral density even though stationarity exists. Furthermore, we investigate the values of α for the aforesaid noise by

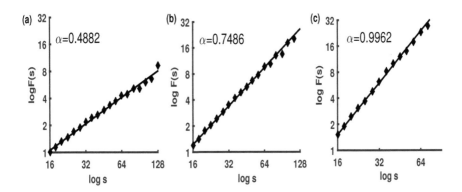

Figure 5.7: (a), (b) and (c) represents $\log_{10} F_{RS}(s)$ vs. $\log_{10} s$ plot for $\beta = 0$, 0.5 and 1 respectively. For each plot, length of the noise are considered 2000. In order to find the HE, straight lines are fitted on the linear region of $\log_{10} F_{RS}(s)$ vs. $\log_{10} s$ plots. On computation, we have chosen scale as 10^s, where $s = [0, 1, 2, 3, 4]$.

DFA. The corresponding results are given in Fig. 5.8c. From Fig. 5.8c, it can be observed that the values of α will always differ for different noise. So, the autocorrelation method and power spectral analysis do not always describe the true characteristics of the signals. Thus, it is better to find the nature of a signal by DFA analysis.

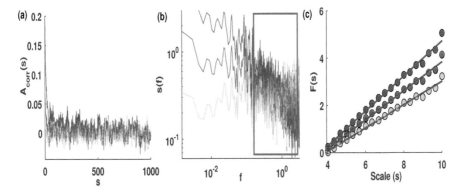

Figure 5.8: (a), (b) and (c) represents $\log s$ vs. $\log F(s)$ plot for $\frac{1}{f^0}$, $\frac{1}{f^{0.5}}$, and $\frac{1}{f}$-noise respectively. For each plot, same length ($L = 2000$) of the time series are considered. The straight lines are fitted on the linear trend of $\log s$ vs. $\log F(s)$ plots. The gradients of the fitted straight lines are considered as the values of α. In each case, scales are chosen as $s = [16, 32, 64, 128, 256, 512, 1024]$.

Note

In DFA, the detrending profile $\tilde{X}(j)$ abruptly changes at the extremum of the segments, since the fitted polynomials in neighbouring segments are sometimes uncorrelated. To adjust the abrupt jump of $\tilde{X}(j)$, windows (in which polynomials are fitted) are taken mutually overlapping. It has been noted that this method takes to much time on computation. Due to this drawback, people have tried to modify the DFA method through: Backward Moving Average (BMA) technique, Centered Moving Average Average (CMA) method, Modified Detrended Fluctuation Analysis (MDFA), Fourier DFA, Empirical mode decomposition, Singular value decomposition, DFA based on high-pass filtering are developed [50].

Sometimes, the a single scaling laws are not able to identify the proper self-similar structure in the STS. In this case, fluctuation is needed to study with different scales and the concept of multifractality is thus proposed. In the following section, we discuss multifractal STS.

5.5 Fluctuation analysis on stationary and non-stationary multifractal time series

To investigate scaling law of a multifractal stationary STS, the standard partition function multifractal formalism based of generalized fluctuation analysis was first proposed. Unfortunately, it fails to reflect the underlying scaling behaviour of the non-stationary STS. This deficiency leds to the development of the wavelet transform modulus maxima (WTMM) method based on wavelet transform.

5.5.1 *Wavelet transform modulus maxima*

For a given continuous STS $\{x(t)\}$, the wavelet transform is generally defined by

$$L_\psi(\tau, s) = \frac{1}{s} \int_{-\infty}^{\infty} x(t)\psi(\frac{t - \tau}{s})dt, \tag{5.17}$$

where $\psi(t)$ represents the mother wavelet from which all the sub-wavelets $\psi_{\tau,s}(t) = \psi(\frac{t-\tau}{s})$ can be calculated.

Similarly, discrete wavelet transform for a STS $\{x(k)\}_{k=1}^{N}$ can be defined as

$$L_\psi(\tau, s) = \frac{1}{s} \sum_{k=1}^{N} x(k)\psi(\frac{k - \tau}{s}) \tag{5.18}$$

The wavelet coefficient $L_\psi(\tau, s)$ depends on both time position τ and frequency scale s. Hence, local fluctuation of the STS can be described in both time and frequency resolution.

In WTMM method, the multifractal scaling coefficient is calculated as follows:

Step-1: Find the position τ_j for which $L_\psi(\tau, s)$ satisfies the condition:

$$|L_\psi(\tau_j - 1, s)| \leq |L_\psi(\tau_j, s)| \leq |L_\psi(\tau_j + 1, s)|,$$

where $j = 1, 2, \ldots, j_{max}$.

This condition gives maxima of $L_\psi(\tau, s)$ over j which is known as modulus maxima. The reason for considering the modulus is

Step-2: Then q-th order fluctuation $Z(q, s)$ is defined by

$$Z(q, s) = \sum_{j=1}^{j_{max}} |L_\psi(\tau_j, s)|^q. \tag{5.19}$$

For increasing s, it can be observed that $Z(q, s)$ obeys the law: $Z(q, s) \sim s^{\tau(q)}$, where $\tau(q)$ is defined by

$$Z_q(s) = \sum_{\nu=1}^{N_s} |X(\nu, s)|^q \sim s^{\tau(q)} \quad \text{for} \quad s > 0. \tag{5.20}$$

The quantity $X(\nu, s)$ is calculated by $X(\nu, s) = \sum_{i=1}^{s} x(\nu s + i)$ for $\nu = 1, 2, \ldots, N_s$, where $N_s = [\frac{N}{s}]$ (N being length of the given STS $\{x(k)\}_{k=1}^{N}$).

Step-3: For each q, $\tau(q)$ is then estimated by fitting a straight line on $\log Z(q, s)$ vs. $\log s$ plot.

The quantity $\tau(q)$ is known as scaling exponent and it characterizes multifractal properties of the STS. In this method the processes of getting the multiscaling exponent is very laborious and takes large computational loops. To overcome this situation, the method of multifractal Detrended Fluctuation Analysis (MF-DFA) was developed. In MF-DFA, fast and reliable results can be obtained compared to WTMM method.

5.5.2 *Multifractal detrended fluctuation analysis*

Multifractal detrended fluctuation analysis (MF-DFA) is basically a combined method of DFA and Generalized Hurst exponent (GHE).

For a given stochastic time series $\{x(k)\}_{k=1}^{N}$, MF-DFA algorithm is described as follows:

Step-1: Calculate the mean $\langle x \rangle$ of the time series $\{x(k)\}_{k=1}^{N}$, where $\langle x \rangle$ is given by

$$\langle x \rangle = \frac{1}{N} \sum_{k=1}^{N} x(k).$$

Check the value of $\langle x \rangle$. If $\langle x \rangle = 0$, then go to step-2. If $\langle x \rangle \neq 0$, then standardize the data to make its mean zero.

Let us consider the general case, i.e; $\langle x \rangle \neq 0$ and the standardize data $\{y(k)\}_{k=1}^{N}$ with $\langle y \rangle = 0$.

Step-2: Construct a profile $X(k)$ from the resultant $\{y(k)\}_{k=1}^{N}$ by

$$X(k) = \sum_{i=1}^{k} [y(i) - \langle y(i) \rangle] \quad (k = 1, 2, \ldots, N). \tag{5.21}$$

Step-3: Define N_s non-overlapping segments A_{ν} $(\nu = 1, 2, \ldots, N_s)$ of length s as given in (5.7).

Step-4: For each segments A_{ν}, fit a local trend $X_{\nu,s}(k)$ (linear or higher order polynomial) on $X(k)$ and subtract from $X(k)$. Then calculate detrended variance $F^2(\nu, s)$ by

$$F^2(\nu, s) = \frac{1}{s} \sum_{k=(s-1)\nu+1}^{s\nu} [X(k) - X_{\nu,s}(k)]^2 \quad (\nu = 1, 2, \ldots, N_s) \tag{5.22}$$

Step-5: Then q-th order fluctuation function F_q is calculated by

$$F_q(s) = \{\frac{1}{N_s} \sum_{\nu=1}^{N_s} [F^2(\nu, s)]^{q/2}\}^{1/q}, \tag{5.23}$$

where q is always taken real valued except zero. Using the above method, it can be seen that 0-th order fluctuation reveals divergent exponents. Instead, logarithmic average approach gives us

$$F_0(s) = \exp\{\frac{1}{2N_s} \sum_{\nu=1}^{N_s} \log[F^2(\nu, s)]\} \sim s^{h(0)}. \tag{5.24}$$

Step-6: Continue the above process with increasing s, a relation between $F_q(s)$ and s can be obtained as $F_q(s) \sim s^{h(q)}$ for the long-term

STS. The exponent $h(q)$ is called scaling exponent or GHE. The value of $h(q)$ is calculated by the slope of the linear regression of $\log F_q(s)$ vs. $\log s$ plot.

Note

For the positive and negative values of q, $h(q)$ describes large and small scale fluctuations respectively. If a time series is monofractal, then $h(q)$ is independent of q. In this case, the scaling behaviour of the variances $F_q(s)$ is identical for all s. On the other hand, it has been observed that $h(q)$ increases with q for the multifractal process. It indicates non-identical $F_q(s)$ for all s. So, the mono-fractality and multifractality of a time series can be identified using GHE. The computation of GHE needs appropriate values of q. If we set $q > 0$, then the segments A_ν with greater fluctuations (segments with relatively high $F^2(\nu, s)$ will give a larger weight in the $F_q(s)$ than that of the segments relatively smaller fluctuation. The opposite holds for $q < 0$. The next important parameter is the degree of the polynomials which are fitted to detrending the fluctuations. Since MF-DFA is suitable for non-stationary process, MF-DFA can be conducted for different polynomials and then decide the best data fit. It has been observed that over fitting leads the value of $F_q(s)$ is close to zero for small values of s. As per selection of the scale is concerned, the smallest scale needs to contain enough elements so that computed local fluctuation can be reliable. In the most of the studies, the minimum scale is taken between 10 and 20. The maximum scale ensures enough elements for computation of the fluctuation function. Most of the studies have been done with $s \not> N/10$.

The multifractal nature can also be investigated by utilizing the concept of Hölder exponent and Legendre transform with GHE [81]. It has been seen that Hölder exponents $\tau(q)$ and GHE $h(q)$ are related by $\tau(q) = qh(q) - 1$. A linear trend in $\tau(q)$ vs. q plot indicates monofractality of the the corresponding time series. On the other hand, nonlinear trend in the same assures existence of multifractality. Example 5.5 illustrates the efficiency of $\tau(q)$ on testing monofractal and multifractal STS.

Example 5.5 *We consider monofractal and multifractal STS in advance, shown in Fig. 5.9a and b respectively. Using (5.24), we first calculate $h(q)$ for both the STS. Then by $\tau(q) = qh(q) - 1$, we investigate $\tau(q)$ for $q \in [-5, 5]$. Figure 5.9c and d shows corresponding $\tau(q)$ vs. q curves for monofractal and multifractal respectively. From the figures, linear and nonlinear trends can be observed in the respective $\tau(q)$ vs. q curves. As linear and nonlinear trends correspond monofractality and multifractality*

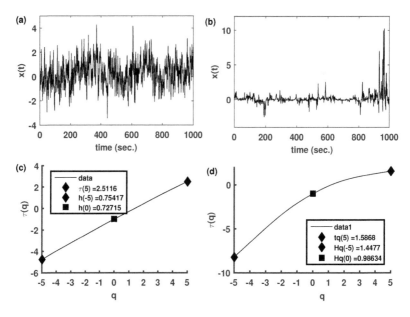

Figure 5.9: (a) and (b) represents power noise $\frac{1}{f^\beta}$ with $\beta = 1.50$ and 1.75 respectively. Each of the time series are of length 3000. (c) represents multifractal spectrum with $\frac{1}{f^{0.5}}$ (solid) and $\frac{1}{f^{0.75}}$ (dotted) respectively. In each computation, scale is chosen $2^N, N = 3, 4, \ldots, 15$ (since degree of the fitted polynomial is taken 1). q is taken as $q \in [-5, 5]$.

respectively, it indicates GHE can characterize underlying fractal nature in a STS.

Furthermore, it has been also established that singularity spectrum $f(\alpha)$ is related to $\tau(q)$ by the relations:

$$\alpha(q) = \frac{d\tau(q)}{dq}, \quad f(\alpha(q)) = q\alpha(q) - \tau(q), \tag{5.25}$$

where $\alpha(q)$ is the Lipschitz-Holder exponent.

We now discuss the effectiveness of $f(\alpha)$ on characterizing two different STS having same statistical properties. Example 5.6 describes the corresponding numerical explanation.

Example 5.6 *We first consider two power noises-$\frac{1}{f^{0.5}}$ and $\frac{1}{f^{0.75}}$ shown in Fig. 5.10a and b respectively. From the figures, it can be investigated that both the STS are statistical equivalents in nature. Using (5.25), We calculate respective $f(\alpha)$s. The corresponding $f(\alpha)$ vs. α curves are shown in*

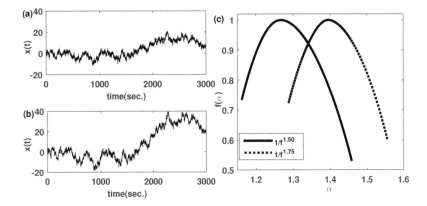

Figure 5.10: (a) and (b) represents power noise $\frac{1}{f^\beta}$ with $\beta = 1.50$ and 1.75 respectively. Each of time series are of length 3000. (c) represents multifractal spectrum with $\frac{1}{f^{0.5}}$ (solid) and $\frac{1}{f^{0.75}}$ (dotted) respectively. In each computation, scale is chosen $2^N, N = 3, 4, \ldots, 15$ (since degree of the fitted polynomial is taken 1). q is taken as $q \in [-5, 5]$.

Fig. 5.10c. From the figure, completely different multifractal spectrums can be observed which indicates separate multifractal structure in the respective STS.

The singularity spectrum of a monofractal signal is represented by a single point in the $f(\alpha)$ plane, whereas the multifractal process reveals a single-humped function. Multifractal analysis can describe the complexity level of a fractal process by quantifying its spectrum. The quantification method was first proposed by Shimizu et al. In this method, the spectrum $f(\alpha)$ is fitted by a regression curve:

$$f(\alpha) = a + b(\alpha - \alpha_0) + c(\alpha - \alpha_0)^2, \tag{5.26}$$

where α_0 is the position of maximum of $f(\alpha)$ and a, b are regression coefficients. The smaller value of α_0 corresponds more correlated and regular process. The coefficient b is known as the asymmetry parameter. Zero value of b indicates symmetric shape. For $b > 0$ and $b < 0$, the respective α vs. $f(\alpha)$ curves will be right and left skewed. Basically dominance of high exponents reveals left skewed nature of $f(\alpha)$ and it indicates a fine structure of the process. On the other hand, the dominance of low exponents indicates more a smooth or regular structure. Further, the width of the spectrum, defined by $W = \alpha_{max} - \alpha_{min}$, measures the degree of the multifractality. Wide ranges indicate the rich structure of the process. α_{max} and α_{min}, known as strongest and weakest Hölder exponent, calculated by considering $f(\alpha_{max}) = 0$ and $f(\alpha_{min}) = 0$. In some cases, quadratic polynomial does not fit on the multifractal spectrum $f(\alpha)$. In this concern, $f(\alpha)$ will be approximated by

$$f(\alpha) = a_1 + a_2(\alpha - \alpha_0) + a_3(\alpha - \alpha_0)^2 + a_4(\alpha - \alpha_0)^3 + a_5(\alpha - \alpha_0)^4. \quad (5.27)$$

To quantify the regression parameters, except α_0 and W, symmetry parameter is calculated in a different way. In fact, symmetry parameter r is defined by

$$r = \frac{\alpha_{max} - \alpha_0}{\alpha_0 - \alpha_{min}}.$$

It has been observed that, the time series possesses symmetry, right skewed and left skewed multifractal spectrum for $r = 1$, $r > 1$ and $r < 1$ respectively.

By applying Legendre transform, the dimension of the fractal process, known as a generalized fractal dimension, can also be calculated. For a stochastic process with Hölder exponent $\tau(q)$, generalized fractal dimension is defined by $D_q = \frac{\tau(q)}{q-1}$. The spectrum of q vs. D_q curve also quantifies the deeper structure of the stochastic process.

5.6   Discussion

This chapter highlights the statistical self-similarity about a univariate time series obtained from the stochastic systems. For an unknown system, it is, therefore, the primary task to check the stochastic nature of the corresponding time series. Several methods have been proposed to identify the stochasticity of a time series [59, 45]. Among them, DVV method [60] is the most robust and effective one. The STS can be obtained in two forms—stationary and non-stationary. A stationary STS shows invariant probability distribution with every time intervals. On the contrary, a variable probability distribution can be observed in the non-stationary STS for every time intervals. We have numerically investigated the stationary as well as non-stationary nature of the time series by auto-correlation methods. The results show the effectiveness of the auto-correlation method.

Self-similar nature is explained for both the stationary and non-stationary stochastic time series. In both the cases, self-similarity can be characterized by the theory of fluctuation in the STS. In order to define fluctuation, different types of statistical measures have been proposed. In fact, these measures correspond to scaling law which characterizes the underlying monofractal and multifractal structure of the

STS. Whether a monofractal STS identifies by a single scaling law, a different scaling law is needed to characterize the multifractal STS.

For a monofractal STS, different methods—auto-correlation function, Fourier based spectrum analysis, Hurst exponent, fluctuation analysis are applied to find the scaling law. However, these are not able to identify the proper law for a non-stationary STS. In this context, the detrended fluctuation analysis is used. On the other hand, multifractality of an STS can be measured by wavelet transform modulus maxima and Generalized Hurst exponent (GHE) methods. Due to the computational inefficiency of the wavelet-based method, we generally apply GHE based fluctuation. It has been seen that GHE can successfully differentiate monofractal and multifractal STS. Moreover, the underlying multifractal structure can be described by the multifractal spectrum which is also computed by GHE. Moreover, two STS which show similar behaviour in time can also be quantified using a polynomial fit technique.

So, monofractal as well as the multifractal stochastic process can be analyzed successfully by fluctuation theory. Hence, the measure of statistical fluctuation can be used for the long-term prediction of dynamics for any STS. It indicates applicability of monofractality and multifractality in the field of stochastic signal analysis.

Chapter 6

Application in Image Processing

6.1 Introduction

Signal processing is a broad branch of electrical engineering. This area is a study about the different signals such as image signals, video signals, voice/sound signals, transmission signals, etc., and their properties.

Image Processing is a sub area of the signal processing. It focuses on the behavior and characteristics of image signals. These image signals are represented into two variants based on the nature of the signals. The categories of the images are analog image and digital image.

An image that manipulates using the analog (electrical) signals is called 'Analog Image' whereas image that is composed by the discrete signals using Sampling and Quantization methods is called 'Digital Image'.

6.1.1 Digital image

Two dimensional discrete function $f(x, y)$ where x and y are spatial coordinates and the amplitude f at (x, y) is defined as digital image. Mathematically Digital image ($\mathbb{R}^2$ space) is represented as a matrix form where x and y coordinate values are mapped with rows and columns of the matrix and the amplitude values are assigned to the cells of the matrix. Digital image of dimension $M \times N$ is represented as $M \times N$ matrix. For each $i = 1, 2, \ldots, M$ and $j = 1, 2, \ldots, N$, (i, j)

presents the location of picture called pixel or picture element. The mapping $f : M \times N \longrightarrow G$, where $G = 0, 1, 2, \ldots, l - 1$, is called the image function or amplitude. The amplitude values are finite and discrete values. The value of f at any pair of spatial coordinates (i, j) describes the intensity or gray level of the image at (i, j).

$$I = \begin{bmatrix} f_{0,0} & f_{0,1} & \cdots & f_{0,N-1} \\ f_{0,1} & f_{0,1} & \cdots & f_{0,N-1} \\ \vdots & \vdots & \vdots & \vdots \\ f_{M-1,0} & f_{M,1} & \cdots & f_{M-1,N-1} \end{bmatrix}.$$

The number of gray levels l is 2^k, where $k \in \mathbb{N}$, usually k gives the bit level. The dynamic range of the intensity of the k bit image is between 0 and $2^k - 1$. For example, 8-bit image has 256 gray levels and the intensity value is between 0 and 255. The number of bits required to store a digital image is $M \times N \times k$.

(a) 128×128 (b) 256×256 (c) 512×512

Figure 6.1: 8-bit camera man image with various sizes.

6.1.2 Digital image processing

The procedure that is processing the digital image by digital computers is called as Digital Image Processing (DIP). DIP finds many applications in a wide-range of domains such as Forensic, Medical, Remote sensing, Manufacturing Industry and consumer electronics. The main functionalities of the DIP are Image Quality improvement using image restoration and image enhancement, extracting the image attributes/features using image segmentation and object recognition and reducing the storage size of the image.

The fundamental steps of DIP are Image acquisition, Image enhancement, Image Restoration, Image Compression, Image Segmentation and Object Recognition (see for more details [56, 93, 63]).

■ Image Acquisition is the process of creating/synthesizing the digital image using different sensors and different modalities. It also involves in the preprocessing methods such as the scaling and shrinking of the digital image.

■ Image Enhancement aims to improve the quality of the image for specific application. It increases the brightness and contrast of the image and also dehazes the image.

■ Image Restoration is a vital preprocessing in DIP. It mainly recovers the digital image from noise and blurring and restores it as its original.

■ The noise is creeping into the image by various sources such as the atmospheric condition, faulty sensors, transmission error, etc. The noise filters are used to suppress the noise in the image.

■ Image Compression is the process of reducing the storage size of the digital image by different compression methods. It resolves the storage issues related to the image.

■ Image Segmentation is the method used to partition the digital image into its regions or objects with respect to the objective of the process. Object recognition is the process that assigns a features/labels to an object based on the descriptors.

This chapter investigates the image segmentation through mutlifractal methods. Further, as a extraction of the mid-sagittal plane from the brains magnetic resonance image, this chapter investigates the possibility of generalized fractal dimensions to measure the asymmetry between hemispheres which separates the parts of human brain.

6.1.3   Image segmentation

Image segmentation is the division of an image into regions or categories with respect to the objective of the process. The image segmentation is carried over by the similarity or dissimilarities intensities in the region of interest. The image segmentation is an application specific/task specific process.

The image segmentation contributes the application in the domains such as defense, medical, remote sensing and industry. In defense, image segmentation is employed for identifying the intruders in the border. In industry, it helps the automation process of manufacturing by detecting defects in the product during their making. In medical, the image segmentation algorithms apply in a wide-range of automate tools. It assists the physicians for the diagnostics of the diseases such as different types of cancers, blocks in blood vessels, etc.

Segmentation adopts three general approaches namely thresholding, edge-based methods and region-based methods. In thresholding, pixels are categorized according to the fixed value/values and the image is partitioned based on the threshold values. In edge-based segmentation, image is segmented by using the different edge filters or edge detection operators such as canny, Laplace, Sobel and Roberts. In region-based segmentation algorithms operate iteratively by grouping together pixels which are neighbours and have similar values and splitting groups of pixels which are dissimilar in value.

6.2 Generalized fractal dimensions

Fractal dimension is insufficient to characterize the object of interest having complex and inhomogeneous scaling properties, since different irregular structures may have same fractal dimensions. Thus, generalized fractal dimensions give more information about the space filling properties than the fractal dimension. Let us review the generalized fractal dimensions followed by fractal dimension.

6.2.1 Monofractal dimensions

Suppose that K is a subset in $\mathbb{R}^n$. The **topological dimension** of K, denoted as $dim_T K$, is inductively defined as follows:

1. $dim_T \emptyset = -1$,

2. The topological dimension of K at a point $p \in K$ is less than or equal to n, written as $dim_T^p K \leq n$, if there exists an arbitrarily small neighborhood of p whose boundaries have topological dimension at most $n-1$,

3. K has topological dimension at most n if it has topological dimension at most n at each of its points p:

$$dim_T K \leq n \Longleftrightarrow dim_T^p K \leq n \text{ for all } p \in K.$$

In addition, $dim_T^p K = \infty$, if the condition (2) does not hold for any $n \in \mathbb{N}$, and $dim_T K = \infty$ if the condition (3) does not hold for any $n \in \mathbb{N}$.

If U is any non-empty subset of n-dimensional Euclidean space $\mathbb{R}^n$, the **_diameter_** of U is defined as $|U| = \sup\{|x - y| : x, y \in U\}$, i.e., the greatest distance apart of any pair of points in U. If $\{U_i\}$ is a countable (or finite) collection of sets of diameter at most δ that cover F, i.e., $K \subset \bigcup_{i=1}^{\infty} U_i$ with $0 < |U_i| \leq \delta$ for each i, we say that $\{U_i\}$ is a δ-cover of K.

Suppose that K is a subset of $\mathbb{R}^n$ and s is a non-negative number. For any $\delta > 0$ we define

$$\mathcal{H}_\delta^s(K) = \inf\left\{ \sum_{i=1}^{\infty} |U_i|^s : \{U_i\} \text{ is a } \delta - \text{cover of } K \right\}.$$

As δ decreases, the class of permissible covers of K is reduced. Therefore, the infimum $\mathcal{H}_\delta^s(K)$ increases, and so approaches a limit as $\delta \to 0$. Thus,

$$\mathcal{H}^s(K) = \lim_{\delta \to 0} \mathcal{H}_\delta^s(K).$$

This limit exists for any subset K of $\mathbb{R}^n$, though the limiting value can be 0 or ∞. We call $\mathcal{H}^s(K)$ as the *s-dimensional Hausdorff measure* of K. Then, the *Hausdorff Dimension* or *Hausdorff-Besicovitch Dimension* of K is defined as,

$$\dim_H(K) = \inf\{s : \mathcal{H}^s(K) = 0\} = \sup\{s : \mathcal{H}^s(K) = \infty\},$$

so that

$$\mathcal{H}^s(K) = \begin{cases} \infty, & \text{if} \quad s < \dim_H(K), \\ 0, & \text{if} \quad s > \dim_H(K). \end{cases}$$

If $s = \dim_H(K)$, then $\mathcal{H}^s(K)$ may be zero or infinite, or may $0 < \mathcal{H}^s(K) < \infty$.

Hausdorff dimension has the advantage of being defined for any set, and is mathematically convenient, as it is based on measures, which are relatively easy to manipulate. A main disadvantage is that the explicit computation of the Hausdorff dimension of a given set K is rather difficult since it involves taking the infimum over covers consisting of balls of radius less than or equal to a given $\epsilon > 0$. A slight simplification is obtained by considering only covers by the balls of radius equal to ϵ. This gives rise to the concepts of box dimension.

Let $K \in \mathscr{H}(X)$ and $\mathcal{N}(K, \epsilon)$ denotes the smallest number of closed balls of radius $\epsilon > 0$ required to cover K. If

$$D_B = \lim_{\epsilon \to 0} \left\{ \frac{ln(\mathcal{N}(K, \epsilon))}{ln(1/\epsilon)} \right\} \tag{6.1}$$

exists, then D_B is called the **box dimension** or **fractal dimension** of K.

6.2.2 Box dimension of image

The box dimension of an image is estimated through the scaling law as

$$N(\epsilon) \propto \frac{1}{\epsilon^{D_B}}$$

$$N(\epsilon) = c\epsilon^{-D_B} \tag{6.2}$$

for $\epsilon \to 0$, where $N(\epsilon)$ denotes the number of boxes of size ϵ which is required to cover the entire intensity surface of the image and c is a constant. In practice, the image includes only the discrete and finite data and therefore the limit of ϵ tents to zero cannot be possible. In order to overcome this difficulty, the fractal dimension of any given image is estimated through the slope of the best fitting line at the point $(-\ln r, \ln N(\epsilon))$, for various values of ϵ.

Let us see the procedure to calculate the $N(\epsilon)$:

$$N(\epsilon) = \sum_{k=1}^{M} kP(k, \epsilon), \tag{6.3}$$

where M is the total number of pixels of the image and $P(k, \epsilon)$ denotes the probability that there are k points within a box of size ϵ, centred about an arbitrary point of the image. The probability $P(k, \epsilon)$ can be estimated by the following relation:

$$P(k, \epsilon) = \frac{n(k, \epsilon)}{N_r} \tag{6.4}$$

where N_r is the number of randomly chosen reference points from the image and $n(k, \epsilon)$ denotes the number of cubes of size r, centred around each reference point, containing the k point of the image.

Differential Box Counting (DBC): Consider an image of size $M \times M$. The domain of the image is partitioned into grids of size $r \times r$. On each grid there is a column of boxes of size $r \times r \times h$, where h is the height of a single box. If the total number of gray levels is G then $G/h = M/r$.

The boxes are numbered sequentially $1, 2, \ldots$. Let the minimum and maximum gray level of the image in $(i,j)^{th}$ grid fall in box number p and q, respectively. Then $n_r(i,j) = q - p + 1$ is the contribution of the $(i,j)^{th}$ grid in $N(\epsilon)$. Taking contributions form all grids, we have

$$N(\epsilon) = \sum_{i,j} n_r(i,j).$$

Because of the differential nature of computing $n_r(i,j)$ the method is called the differential box counting (DBC) approach. Calculating $N(\epsilon)$ in this manner gives a better approximation to the boxes intersecting the image intensity surface, especially when there are sharp gray level variations in neighbouring pixels.

Relative Differential Box Counting (RDBC): A modification of the DBC, called the relative differential box-counting (RDBC) method, was proposed. According to this method, $N(\epsilon)$ is obtained by the following equation:

$$N(\epsilon) = \sum_{i,j} \lceil (kd_r(i,j)/r) \rceil,$$

where $d_r(i,j)$ denotes the difference between the maximum and the minimum gray level of the image in the grid (i,j), $k = M/G$ and $\lceil (x) \rceil$ denotes for the ceiling function of x, i.e., the smallest integer which is greater than or equal to x.

Correlation Algorithm: A very popular way to compute the dimension is to use the correlation algorithm, which estimates dimension based on the statistics of pairwise distances. According to this algorithm the dimension is defined as

$$\nu = \lim_{r \to 0} \frac{\ln C(r)}{\ln r},$$

where $C(r)$ is the correlation integral given by

$$C(r) = \frac{\text{Number of distances less than } r}{\text{Number of distances altogether}}.$$

The correlation algorithm provides a particularly elegant formulation and simultaneously has the substantial advantage that the function $C(r)$ is approximated even for r as small as the minimum interpoint distance. For an image with M pixels, $C(M,r)$ has a dynamic range of $O(M^2)$. Logarithmically speaking, this range is twice that available in the box-counting method (see, [46, 72, 84]).

6.2.3 *Multifractal dimension*

Alfred Renyi introduced the measure to quantify the uncertainty or randomness of a given system. It has a vital role in information theory. Given probabilities p_i with $\sum_{i=1}^{N} p_i = 1$, the Renyi entropy of order q is given by

$$RE_q = \frac{1}{1-q} ln \sum_{i=1}^{N} p_i^q,$$

where $q \geq 0$ and $q \neq 1$. At $q = 1$ the value of RE_q is potentially undefined as it generates the indeterminate form, otherwise RE_q values are decreasing as a function of q.

If $q \to 1$, then $RE_q \to RE_1$ which is defined by

$$RE_1 = -ln \sum_{i=1}^{N} p_i ln p_i.$$

RE_1 is called **Shannon entropy** [34, 43]. The ***Renyi Fractal Dimensions*** *or* ***Generalized Fractal Dimensions*** (GFD) of order $q \in (-\infty, \infty)$ is defined, in terms of generalized Renyi Entropy, as

$$D_q = \lim_{r \to 0} \frac{1}{q-1} \frac{\ln\left(\sum_{i=1}^{N} p_i^q\right)}{\ln_2 r} \tag{6.5}$$

where p_i is probability distribution. As $q \longrightarrow 1$, D_q converges to D_1, which is given by

$$D_1 = \lim_{r \to 0} \frac{\sum_{i=1}^{N} p_i \ln p_i}{\ln r}, \tag{6.6}$$

where D_1 is the **information dimension** and D_q is a monotonically decreasing function of q such that $D_0 \geq D_1 \geq D_2$. Here D_0 and D_2 denotes the fractal dimension and correlation dimension respectively.

Multifractal Spectra: The multifractal spectra $f(\alpha(q))$ is defined as

$$f(q) = \lim_{r \to 0} \frac{\sum_{i=1}^{N(r)} \mu_i(q,r) \ln(\mu_i(q,r))}{\ln r}, \tag{6.7}$$

$$\alpha(q) = \lim_{r \to 0} \frac{\sum_{i=1}^{N(r)} \mu_i(q,r) \ln(P_i(q,r))}{\ln r}, \tag{6.8}$$

the normalized measures μ_i is defined, in terms of probability P_i, as

$$\mu_i(q,r) = \frac{[P_i(r)]^q}{\sum_{i=1}^{N(r)} [P_i(r)]^q}, \tag{6.9}$$

here $N(r)$ is the number of boxes required to cover the object with box size r and P_i is a probability of ith box of size r. Generally P_i is defined as

$$P_i = \frac{area\ of\ i^{th}\ part}{total\ area}.$$

Some properties of multifractal spectra as follows:

- The spectrum of $f(\alpha)$ is concave,

- $f(\alpha)$ has a single inflection point at q=0,

- At q=0, f achieves maximum. $f(\alpha(0)) = D_0$ refers the fractal dimension,

- In case of monofractal analysis the spectrum of $f(\alpha)$ is reduced to a single point.

(a) Cameraman   (b) Lena   (c) Mandril

(d) Peppers   (e) Pirate

Figure 6.2: Standard 8 bit images used for segmentation.

Table 6.1: Generalized fractal dimensions of standard images.

Image	D_0	D_1	D_2
Cameraman	1.1959	1.1881	1.1855
Lena	1.2434	1.2398	1.2379
Mandrill	1.2489	1.2480	1.2471
Peppers	1.2465	1.2433	1.2407
Pirate	1.2412	1.2341	1.2294

6.3   Image thresholding

In thresholding, pixels are categorized according to the fixed value (threshold value) and the image is divided based on the threshold value. In this section, we partition the image based on the multifractal dimension as a threshold value.

6.3.1   *Multifractal dimensions: A threshold measure*

After the sampling and quantization process, image can be represented as a matrix form. Let M, N be the finite subsets of natural numbers and $G = \{0, 1, 2, ..., k-1\}$ be a set of positive integers to denote the gray levels of k-bit. Then, an $M \times N$ dimensional image can be defined as a function $f : M \times N \longrightarrow G$ by $f(x, y) = i, i \in G$. Assume that $t \in G$ is a optimal thresholding value and $B = \{0, 1\}$ be the binary gray levels in G. Then the thresholding image can be defined as a mapping $T : M \times N \longrightarrow B$, such that

$$T(x, y) = \begin{cases} 0 & if \ f(x, y) \leq T \\ 1 & if \ f(x, y) > T \end{cases}$$

A gray scale image can be described in term of a mass distribution. We assume that the total intensity as a finite mass scattered to the whole image so that the white areas have high density and black areas have low density. we have to determine threshold (t) value as alone from the gray level of each pixel through following steps.

Step 1: Read the input MRI.

Step 2: Let N be the number of boxes to cover the image with box size r.

Step 3: The probability p_i for i^{th} box of size r in the image is defined as,

$$p_i = \frac{X_i}{X}$$

where X_i is the intensity value of the image in the corresponding i^{th} box of size r and X is the total intensity value image.

Step 4: Estimate the value of N.

Step 5: Fix q, calculate D_q as defined in Eq. 6.5 for various $r \to 0$.

Step 6: Repeat step 5 for various $q \in (-150, 150)$, we found D_q for each intensity level of the given input image using Eq. 6.5.

Step 7: Find the median value of D_q's and fix corresponding intensity level as the optimal threshold, $t = med(D_q)$.

Step 8: Based on threshold value, the image is partitioning as foreground and background regions. Binary image B(x,y) generated from the original image $f(x, y)$ as

$$B(x, y) = \begin{cases} 0 & if \ f(x, y) \le t \\ 1 & if \ f(x, y) > t \end{cases}$$

Step 9: Mask the input image by generated binary image.

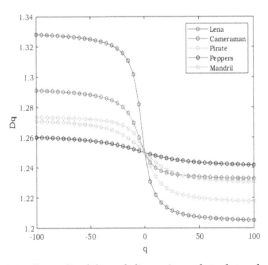

Figure 6.3: Generalized fractal dimensions plot of standard images.

6.4 Performance analysis

6.4.1 Evaluation measure for quantitative analysis

The region nonuniformity measure pronounces the inherent quality of the segmented region. Assume that $f(x, y)$ is given gray scale image then the region nonuniformity measure RNU is defined as

$$RNU = \frac{|FG| \times Var(FG)}{|FG + BG| \times Var(f)},$$

where FG is foreground image pixels, BG is background image pixels, Variance of the whole image is denoted by $Var(f)$ and $Var(FG)$ represents the variance of the foreground region of the given image $f(x, y)$, $|.|$ cardinality of the given object. Good segmented images have a RNU value close to 0.

6.4.2 Human visual perception

The proposed method uses the generalized fractal dimensions as a thresholding measure to segment the gray scale images. Let us discuss the result of the proposed method for foreground extraction. As mentioned in Chapter 2, one way to define a fractal is with non-integer Hausdorff dimension, which exceeds its topological dimension. According to this definition of fractal, In Eq. (6.5), choose $q = 0$ then it provides the fractal dimension D_0. Table 6.1 reveals the fractal dimension D_0 of standard images given in Fig. 6.2 and it lies between 0 and 1. However, the topological dimension of gray scale image is 1 as they are considered as objects in the Euclidean plane. Further, generalized fractal dimensions of standard images depicted in Fig. 6.3 where q values consider from -100 to 100. Threshold value of standard images obtained by the proposed method and Otsu method are elucidated in Table 6.2, Table 6.3 respectively. Further, Region nonuniformity of these two methods are also given in Table 6.2, Table 6.3. Extracted foreground images by proposed method are depicted in Fig. 6.4 and the foreground of standard images attained through Otsu method are given in Fig. 6.5.

The objective of the proposed algorithm is to develop the precise technique to find the optimal threshold value and thereby extract the foreground of given gray scale image. Moreover, the developed method has to provide the better performance on irregular gray level distribution. In order to analyze the efficiency of the proposed method, it has been compared with the state-of-the-art method namely the Otsu method.

Table 6.2: Threshold value and Region nonuniformity of standard images.

Image	Threshold value	Region nonuniformity
Cameraman	137	0.0305
Lena	133	0.1378
Mandrill	130	0.1271
Peppers	122	0.0973
Pirate	131	0.1098

Table 6.3: Threshold value and Region nonuniformity by Otus method.

Image	Threshold value	Region nonuniformity
Cameraman	86	0.0466
Lena	116	0.1374
Mandrill	126	0.1372
Peppers	106	0.1267
Pirate	115	0.1078

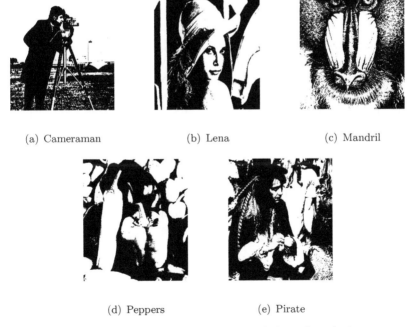

(a) Cameraman   (b) Lena   (c) Mandril

(d) Peppers   (e) Pirate

Figure 6.4: Segmented images by multifractal method.

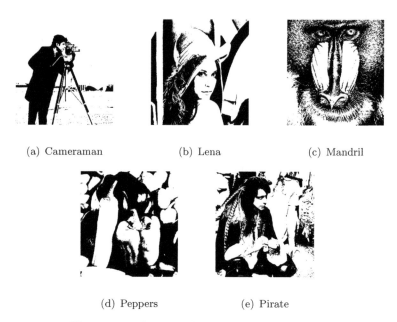

(a) Cameraman   (b) Lena   (c) Mandril

(d) Peppers   (e) Pirate

Figure 6.5: Segmented images by Otsu method.

The performance of the developed algorithm on standard images is evaluated by quantitative metric namely Region nonuniformity. These values are recorded in Table 6.2 and Table 6.3, which exhibits the comparison between the extracted foreground by the proposed method and the Otsu method. It is evident that the multifractal based method gives Region nonuniformity value significantly close to 1 which is better than Otsu method Region nonuniformity value.

In order for the human visual perception analysis, extracted foreground of standard images dataset through developed method and Otsu method are depicted in Fig. 6.4 and Fig. 6.5. This comparison depicts the out performance of the proposed method for extraction of the foreground from the specified images dataset.

6.5   Medical image processing

Medical image diagnosis is one of the emerging fields which is dealing with high-technology and modern instrumentation that plays a vital role in diagnosis, analysis, and treatment. One among the precise techniques of diagnosis processes is the segmentation of the medical images. The aim of medical image segmentation lies in extracting the region of interest (RoI), in order to study the intrinsic anatomical structure and thereby estimating the severity of abnormalities that helps for the prognosis of the disease. This situation gives rise to the development of efficient computing techniques for segmentation and analysis. Brain magnetic resonance image segmentation is an essential step to classify the anatomical areas of interest for diagnosis, treatment, medical planning of tumours, infarct and stroke.

The human brain is partitioned into two hemispheres, left and right respectively by a tissue layer called inter-hemispheric fissure (IF) which is filled with cerebrospinal fluid (CSF). This partition is done by Mid-sagittal Plane (MSP) along with IF, which is generally estimated as a plane passing through those hemispheres. The development of algorithms for the automated detection of MSP has wide applications in the analysis of brain tumours through Computer Assisted Diagnosis (CAD). As brain tissues have complex structures, fractal analysis assumes that the object of interest can be described by a single fractal dimension alone. Thus, multifractal analysis provides us more information about the space filling properties than the fractal dimension D_0. In order to overcome this limitation, the proposed algorithm aims to detect and segment MSP precisely from a brain image by exploring the potential of generalized fractal dimensions along with its multifractal spectra.

Many algorithms for the detection of MSP have been reported on the literature in which most of the works are concentrated on image intensity in order to detect the symmetry plane [51, 78, 79, 82]. Unfortunately, most of the methods lack in analyzing the image slices by considering the texture of the image. The human brain possesses a complex geometric structure which cannot be described only through Euclidean geometry. As brain tissues possess to have self-similarity structures, fractal based analysis would be prompt to detect MSP from a brain image. Fractal dimension analyzes the irregularity of an object with homogeneous scaling properties. The concept of fractal dimensions can be realistic in the measurement and categorization of shape and texture. It is a fact that only a handful of works have been depicted using fractal analysis for the detection of MSP. Jayasuriya et al. [82] have proposed a method for identifying the symmetry plane in three-dimensional brain MRI images, based on the analysis using fractal dimension and lacunarity. However, fractal dimension is insufficient to characterize the object of interest having a complex and inhomogeneous scaling properties, since different irregular structures may have same fractal dimension. Lacunarity measurements includes significantly to the description of an object with a known fractal dimension, which describes the empty space around the object, and thus it relates to how the object fills the space. As brain tissues possess to have complex structures, fractal analysis assumes that object of interest can be described by a single fractal dimension alone. Thus, multifractal analysis provides more information about the space filling properties than fractal dimension D_0 (see for details, [14, 15, 16, 84, 89, 96]). In order to overcome this limitation, the present work aims to detect and segment MSP precisely from a brain image by exploring the potential of generalized fractal dimensions along with its multifractal spectra.

6.6 Mid-sagittal plane detection

The mid-sagittal plane divides the brain into two similar hemispheres, namely right and left hemispheres. Each hemisphere should have similar Generalized Fractal Dimensions (GFD), hence the difference between GFD of left hemisphere and GFD of right hemisphere should be minimum at the MSP. Hence, the present method defines symmetry measure, in terms of GFD, as

$$\gamma = \frac{|GFD_R - GFD_L|}{GFD_W}, \tag{6.10}$$

where GFD_R, GFD_L and GFD_W denote the generalized fractal dimensions of right hemisphere, left hemisphere and whole image of the MRI respectively. When γ tends to zero then the MSP is optimal.

For the input MRI I, the proposed method initially calculates the GFD_W for the whole image I. Based on the intensity levels, the extracted RoI is partitioned into left and right regions respectively for which GFD is calculated accordingly as (GFD_L, GFD_R). Now the symmetry measure is calculated using the values of GFD_L, GFD_R and GFD_W. Considering the optimal intensity using minimum symmetry measure, the sagittal plane is found out. Apply multifractal spectra on s to detect the final mid-sagittal plane from input MRI. The step-by-step algorithm for the above said methodology is given as below.

Algorithm: Mid-sagittal Plane Detection from MRI

Phase I: Computation of Generalized Fractal Dimension of MRI

Step 1: Read a brain MRI I.

Step 2: Let N be the number of boxes required to cover the image I with box size r.

Step 3: The probability P_i for i^{th} box of size r in the image is defined as,

$$P_i = \frac{X_i}{X},$$

where X_i is the intensity value of the image in the corresponding i^{th} box of size r and $X = \sum_{i=1}^{N} X_i$.

Step 4: Fixing q and varying r, compute D_q using Eq. (6.5)

Begin
  $for\ i : 1 \longrightarrow r$
  $for\ j : 1 \longrightarrow r$
  $mass(i,j) = sum(sum(I(i*N - (N-1) : i*N, j*N - (N-1) :$
  $j*N)));$
  $p(i,j) = mass(i,j)/X;$
  $if\ q \neq 1$
  $p_q(i,j) = [p(i,j)]^q;$
  else
    $p_q(i,j) = q \times [p(i,j)];$
    $pi \leftarrow sum(sum(p_q));$
    $D(q) \leftarrow \ln(pi)/\ln(r);$
End

Step 5: Repeat step 4 for various $q \in (-\infty, \infty)$, to acquire $GFD_W \leftarrow$ GFD of I.

Phase II: Detection of Mid-sagittal Plane

Step 6: Extract the initial sagittal region Δ from the image I by removing/subtracting the extreme 20% of region on both sides.

Step 7: Based on each intensity of Δ, partition Δ into left Δ and right Δ respectively.

Step 8: Estimate GFD of left Δ and right Δ to acquire GFD_L and GFD_R respectively.

Step 9: Estimate the symmetric measure $\gamma = \frac{|GFD_L - GFD_R|}{GFD_W}$.

Step 10: Estimate c^* to be the optimal c for which γ is minimum.

Step 11: Consider the RoI centered on c^*, and select sagittal plane s within the RoI centered on $[c^* \pm \delta]$, where δ as 0.5 mm.

Step 12: Apply multifractal spectra on s using Eq. (6.7) and Eq. (6.8) to acquire I_{MF}.

Step 13: Compute the average multifractal spectra values λ_s of I_{MF} as

Begin $for\ i : 1 \longrightarrow N$

$$\mu_i(q, r) = \frac{P_i}{\sum_{j=1}^{N} P_j^q};\ I_1 = \frac{\sum_{i=1}^{N} P_i^{q-1}}{N};$$

$$\tau(q) \leftarrow \text{imgradient}(polyfit(I_1, q, 1);$$

$$I_2 = \sum_{i=1}^{N} \mu_i \times log P_i;$$

$$\alpha(q) \leftarrow \text{imgradient}(polyfit(I_2, q, 1);$$

$$I_3 = \sum_{i=1}^{N} \mu_i \times log \mu_i;$$

$$f(\alpha(q)) \leftarrow \text{imgradient}(polyfit(I_3, q, 1);$$

End

compute $\lambda_s \leftarrow \dfrac{\sum_{q \in s} f(\alpha(q))}{\sum_{q \in s} q}.$

Step 14: Find the optimal sagittal plane s_0 that gives the $\Delta\gamma$ value for γ from ROI of MRI planes, where $\Delta\gamma = \gamma_{max} - \gamma_{min}$.

Step 15: Output s_0, MSP.

Step 16: Stop.

In the developed algorithm, Phase I elucidated the computational procedure of generalized fractal dimensions for the input MRI. In order to obtain an accurate estimation of similarity between the left and right hemispheres, the symmetry measure is defined in terms of GFD as illustrated in **Step 9**. Hence, the proposed algorithm outperforms on asymmetrical or pathological images.

6.6.1 Description of experimental MRI data

The MRI datasets are taken from the Centre for Morphometric Analysis at Massachusetts General Hospital and is available at http://www.cma.mgh.harvard. edu/ibsr/ [76] and BrainWeb at http://www.bic.mni.mcgill.ca/brainweb/ [77]. In order to evaluate the performance of the developed method, normal and pathological MRI samples are used, which are given in Table 6.4. Further, simulated data from BrainWeb were used to acquire the accuracy of proposed method by comparing it with the ground truth line. The data volumes have simulated using T_1 and T_2 weighted sequences with slice thicknesses 3 mm, 5 mm, 7 mm and intensity non-uniformities (INU) 20%, 40% along with the noise levels 5%, 7%, 9%. These datasets are available for showing in coronal orthogonal view, which is illustrated in Fig. 6.6 and Fig. 6.7.

6.6.2 Performance evaluation metrics

A quantitative measure is used to acquire the accuracy of the proposed method by comparing it with the ground truth MSPs which are marked manually, which is being provided by Internet Brain Segmentation Repository (IBSR) and BrainWeb data. This is done with the measure of the values of Yaw angle error and Roll angle error (for more details, [51]). The computational details of evaluation metrics are explained as follows.

MSP can be described, in terms of coordinate system, as $aX + bY + cZ + d = 0$, where (a, b, c) is the scaling vector and $\dfrac{d}{\sqrt{a^2 + b^2 + c^2}}$ is vertical distance from the origin. Then the Yaw angle (ϕ_y) and Roll angle(ϕ_r) are defined as

$$\phi_y = arctan\frac{b}{a} \tag{6.11}$$

$$\phi_r = arctan\frac{-c}{\sqrt{a^2 + b^2}}. \tag{6.12}$$

Table 6.4: Brain MRI datasets for evaluation of the proposed method.

Pathology	#Sample	Modality	Matrix	Orientation
Infarct	23	T_1, T_2	$256 \times 256 \times 173$	sagittal
Tumour	35	T_1, T_2	$256 \times 256 \times 124$	sagittal, axial
Stroke	24	T_1	$256 \times 256 \times 76$	sagittal, axial
Normal	33	T_1	$256 \times 256 \times 256$	sagittal, coronal

Slice thickness	5% Noise	7% Noise	9% Noise
3 mm	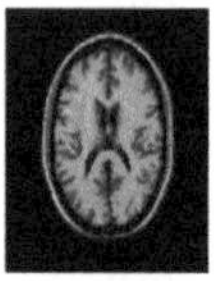	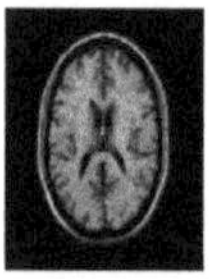	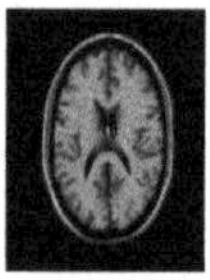
5 mm	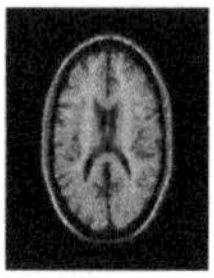	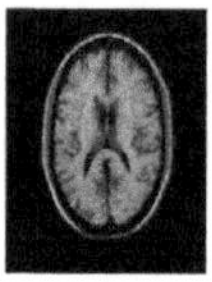	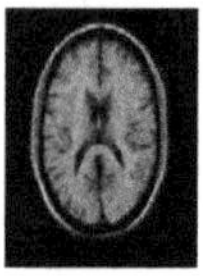
7 mm	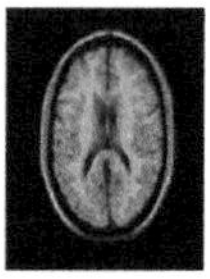	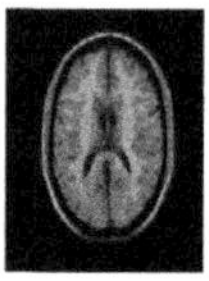	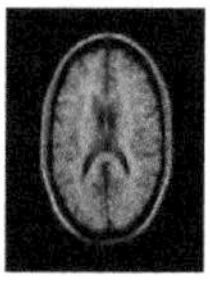

Figure 6.6: Simulated brain MRI with intensity non-uniformity 20% with the different noise levels.

The Yaw angle error ϕ'_y and Roll angle error ϕ'_r estimated as a difference between MSP detected by the proposed method and the ground truth MSP.

The angular deviation θ is calculated between ground truth line and MSP estimated by the proposed method. Furthermore, the average deviation of the distance (d, in pixels) is estimated from the upper and lower endpoints between the estimated MSPs and the ground truth lines as

$$d = \frac{\sqrt{(a - x)^2 + (b - y)^2} + \sqrt{(a' - x')^2 + (b' - y')^2}}{2}, \qquad (6.13)$$

where $(a, b), (a', b')$ denotes the upper and lower endpoints of the MSP estimated by the proposed method. Also $(x, y), (x', y')$ denotes the upper and lower endpoints of the ground truth MSP [82].

6.6.3 Results and discussions

Jayasuriya et al. proposed the fractal based method for extracting the symmetry plane from brain MRI. The limitation of their algorithm is

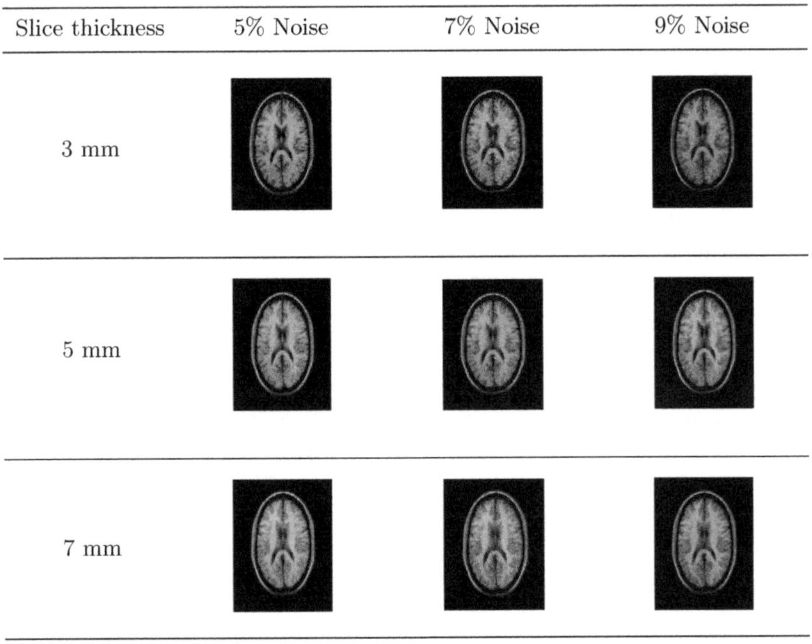

Slice thickness	5% Noise	7% Noise	9% Noise
3 mm			
5 mm			
7 mm			

Figure 6.7: Simulated brain MRI with intensity non-uniformity 40% with the different noise levels.

that it still requires a certain degree of symmetry between the left and right hemispheres. Therefore, it may not provide accurate results on images with severe global asymmetry, like substantial hemispheric removal. This limitation couldn't be successfully defeated for all correlation based techniques. In order to overcome this problem, the proposed method uses the generalized fractal dimensions as a symmetry measure and multifractal spectra for the accurate estimation of MSP from brain MRI samples. The results of the proposed method for MSP extraction are presented here.

The fractal is defined as a set with non-integer Hausdorff dimension, which exceeds its topological dimension. In Eq. (6.5), choose $q = 0$ then it provides the fractal dimension D_0. In Table 6.5 and Table 6.6, the fractal dimension value D_0 which lies between 0 and 1, further digital MRI datasets used in the proposed method are subsets of the Euclidean plane, which have the topological dimension 1. Hence, these images are fractal objects, which have irregular distribution of the gray levels although they have a self-similar structure.

In Table 6.5, the tabulated values of generalized fractal dimensions D_0, D_1 and D_2 of simulated MRI data as shown in Fig. 6.6, which are computed by fixing the intensity non-uniformity value 20% with slice thickness of 3 mm, 5 mm, 7 mm and varying the noise levels by 5%, 7%, 9%. The multifractal strength $\triangle \alpha$ is estimated by the difference between minimum and maximum values of the multifactal spectra. The values enlisted in the Table 6.6 are computed for 40% INU with same slice thickness and varying noise levels as mentioned above of MRI data as depicted in Fig. 6.7.

Figure 6.8 depicts the GFD spectra for q values which lie between -100 to 100 and the multifractal spectra for RoI centered at optimal intensity c^* obtained from **Step 10** in the proposed algorithm. Figure 6.8 and Fig. 6.9 explains the multifractal spectra of brain MRI datasets with INU 20% and INU 40% which respectively is convex with respect to varying slice thicknesses and noise levels with a single inflection point at $\alpha(0)$, which provides the fractal dimension D_0. Further, Fig. 6.10 and Fig. 6.11 depicts the extracted MSP by the developed method for MRIs in the Fig. 6.6 and Fig. 6.7 respectively.

Table 6.7 presents the mean and standard deviation (SD) of angular deviation θ between ground truth lines and MSP estimated by the proposed method, here θ is computed for datasets tabulated in Table 6.4. Further, Table 6.7 provides the mean and standard deviation (SD) of the average deviation of distance d between lower and upper endpoints of estimated MSP and ground truth MSP, which is determined from Eq. (6.13).

The MSP extracted from the proposed method is compared with the three state-of-the-art methods presented by Liu et al. [78], Ruppert et al. [79] and Zhang et al. [65]. In order for the human visual perception analysis, extracted MSP of a pathological and normal brain MRI

Table 6.5: Analysis of multifractal strength $\triangle \alpha$, maximum value of the multifractal spectrum α_{max} and generalized fractal dimensions of brain MRI with INU 20% with slice thickness.

MRI Slice Thickness	INU 20% with Noise Levels	α_{max}	$\triangle \alpha$	D_0	D_1	D_2
3* 3 mm	5% Noise	2.9666	1.1829	1.9205	1.8229	1.8208
	7% Noise	2.9406	1.1546	1.8275	1.8230	1.8210
	9% Noise	2.6786	0.8828	1.8297	1.8233	1.8216
3*5 mm	5% Noise	2.8899	1.1041	1.8273	1.8233	1.8217
	7% Noise	3.0290	1.2480	1.8286	1.8243	1.8229
	9% Noise	2.7538	0.9554	1.8253	1.2834	1.8218
3* 7 mm	5% Noise	3.0293	1.2426	1.8267	1.8242	1.8232
	7% Noise	3.0343	1.2445	1.8294	1.8239	1.8227
	9% Noise	2.7204	0.9201	1.9204	1.8244	1.8235

Table 6.6: Analysis of multifractal strength $\triangle\alpha$, maximum value of the multifractal spectrum α_{max} and generalized fractal dimensions of brain MRI with INU 40% with slice thickness.

MRI Slice Thickness	INU 40% with Noise Levels	α_{max}	$\triangle\alpha$	D_0	D_1	D_2
3* 3 mm	5% Noise	2.8756	1.0937	1.8264	1.8243	1.8233
	7% Noise	2.9488	1.1653	1.8290	1.8241	1.8230
	9% Noise	2.8816	1.0924	1.8253	1.8244	1.8236
3* 5 mm	5% Noise	2.7310	0.9468	1.8257	1.8232	1.8214
	7% Noise	3.0570	1.2697	1.8252	1.8231	1.8212
	9% Noise	2.8060	1.0092	1.8253	1.8233	1.8216
3* 7 mm	5% Noise	3.0755	1.2892	1.8284	1.8243	1.8234
	7% Noise	2.9663	1.1762	1.8253	1.8243	1.8231
	9% Noise	2.8170	1.0190	1.8261	1.8242	1.8232

Table 6.7: Comparison of proposed method with ground truth MSP through angular deviation θ and average deviation of distance d.

2*Dimension	θ (in degree)		d (in pixels)	
	Mean	Standard Deviation	Mean	Standard Deviation
D_0	0.91	0.21	0.72	0.19
D_1	0.89	0.25	0.69	0.23
D_2	0.85	0.27	0.67	0.27

datasets from aforementioned methods and proposed method are depicted in Fig. 6.12. Moreover, Yaw angle error ϕ'_y and Roll angle error ϕ'_r are computed for the proposed method and the existing methods using Eq. (6.11) and Eq. (6.12). Table 6.8 provides the mean and standard deviation of those ϕ'_y and ϕ'_r estimated for the considered datasets illustrated in Table 6.4.

The objective of the present study is to develop the precise technique for appropriate estimation of the mid-sagittal plane from normal and pathological brain MRI. Moreover, the developed method has provided better performance on irregular/asymmetrical MRI structure. In order to analyze the stability of the proposed method, the Gaussian noise is to be added along with intensity non-uniformity levels of 20% and 40% in MRI at the time of simulation from BrainWeb database. Besides, pathological images are also used to evaluate the efficiency of the proposed method.

From Table 6.8, the mean and SD values of Yaw angle error ϕ'_y and Roll angle error ϕ'_r of the proposed method is significantly minimum among the other three state-of-the-art methods. Also, Table 6.8 exhibits

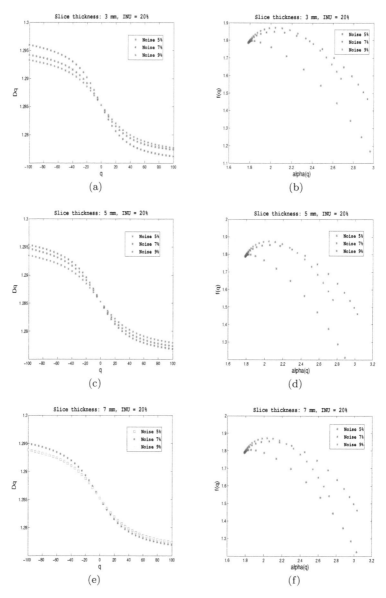

Figure 6.8: Comparative analysis of multifractal spectra and generalized fractal dimensions spectra of simulated brain MRI with INU 20%, noise levels 5%, 7%, 9% and various slice thickness.

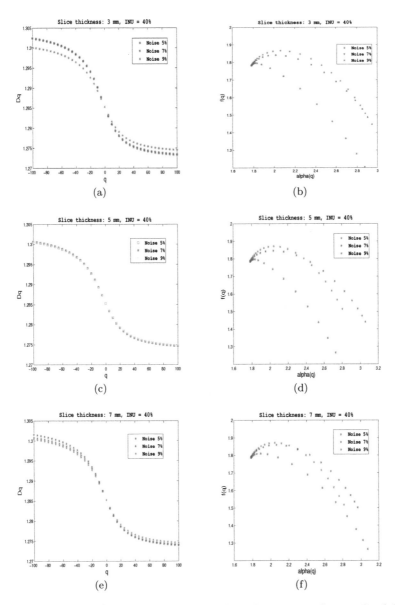

Figure 6.9: Comparative analysis of multifractal spectra and generalized fractal dimensions spectra of simulated brain MRI with INU 40%, noise levels 5%, 7%, 9% and various slice thickness.

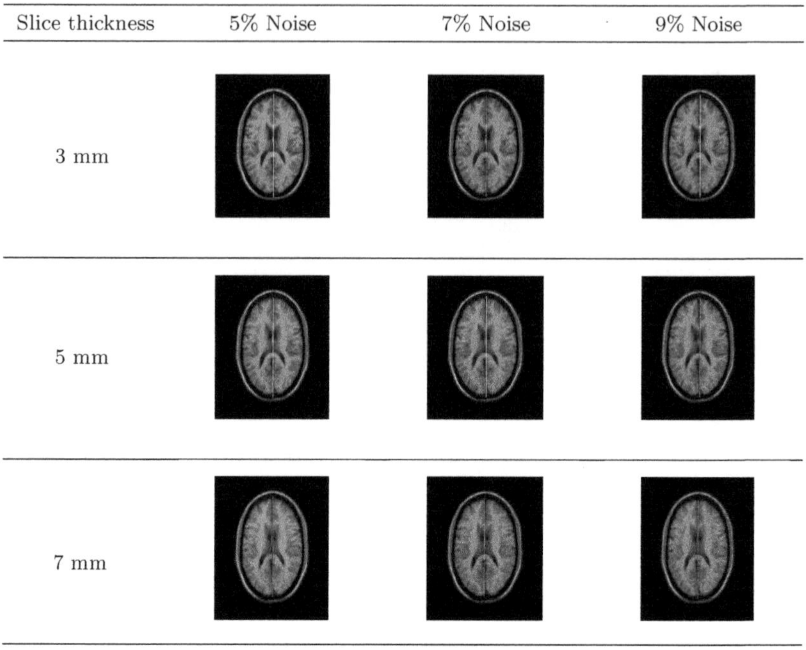

Slice thickness	5% Noise	7% Noise	9% Noise
3 mm			
5 mm			
7 mm			

Figure 6.10: Extracted MSP from the simulated MRI brain volume as shown in Fig. 6.6.

the efficiency of the method by using multifractal techniques such as generalized fractal dimensions and multifractal spectra. In addition, Table 6.7 and Fig. 6.12 depict the superior performance of proposed method for extraction of the MSP on pathological MRI such as tumour, infarct and stroke.

The performance of the developed method on pathology and normal MRIs is evaluated by another quantitative metric namely angular deviation θ and average deviation of distance d. These values are recorded in Table 6.7, which exhibits the comparison between the extracted MSP and the ground truth line. It is evident that the multifractal based method is significantly close to the ground truth MSP via angle and pixel distance. The performance of the developed method on a simulated MRI obtained from the BrainWeb database is given in Fig. 6.10 and Fig. 6.11 for human visual perception.

Table 6.8: Comparison using Yaw angle error ϕ'_y and Roll angle error ϕ'_r of MSP.

| Type | Zhang and Hu | | | | Ruppert et al. | | | | Liu et al. | | | | Proposed Method | | | |
| | ϕ'_y | | ϕ'_r | | ϕ'_y | | ϕ'_r | | ϕ'_y | | ϕ'_r | | ϕ'_y | | ϕ'_r | |
	Mean	SD	Mean	SD	Mean	SD	Mean	SD	Mean	SD	Mean	SD	Mean	SD	Mean	SD
Infarct	2.432	0.317	2.323	0.375	2.217	0.271	2.243	0.295	1.877	0.217	1.813	0.234	**0.913**	**0.153**	**0.921**	**0.178**
Tumour	2.340	0.312	2.317	0.369	2.204	0.274	2.043	0.291	1.873	0.213	1.795	0.227	**0.897**	**0.149**	**0.917**	**0.163**
Stroke	2.351	0.309	2.371	0.371	2.301	0.269	1.976	0.283	1.732	0.209	1.703	0.229	**0.907**	**0.143**	**0.873**	**0.159**
Normal	2.209	0.315	2.246	0.367	1.989	0.267	1.896	0.280	1.708	0.211	1.702	0.223	**0.793**	**0.148**	**0.781**	**0.152**

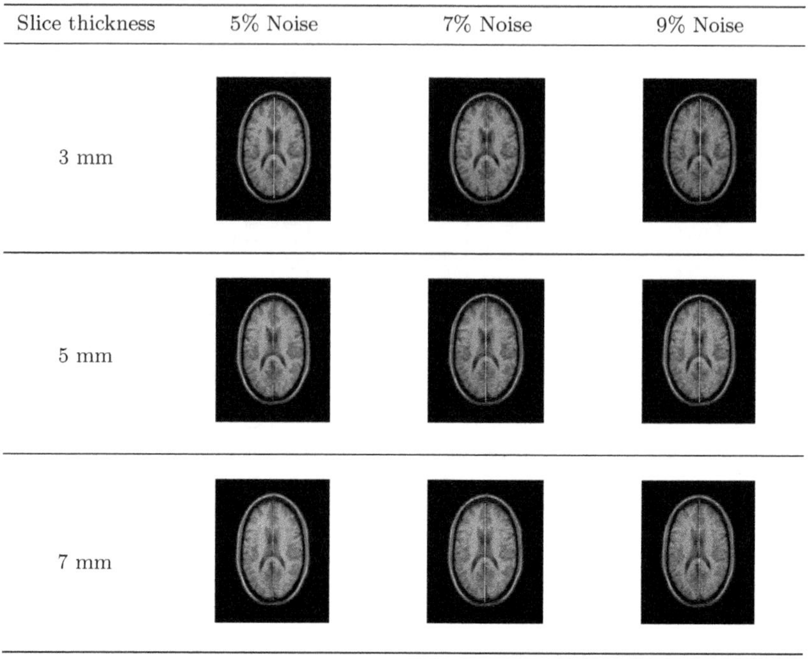

Slice thickness	5% Noise	7% Noise	9% Noise
3 mm			
5 mm			
7 mm			

Figure 6.11: Extracted MSP from the simulated MRI brain volume as shown in Fig. 6.7.

In order to compare the proposed method along with the existing methods for MSP extraction by evaluation metrics and human visual perception, the developed technique has following merits: (1) Generalized fractal dimensions and multifractal spectra would be comfortable enough to characterize the noise behaviour of MRI, hence the developed method has a good tolerance of noise. (2) The algorithm works so well on various intensities non-uniformity, and slice thicknesses of pathological images. (3) GFD is used for symmetry measure, so that it computes the accurate symmetry between the left and right hemispheres.

Degrees of symmetry between the left and right hemispheres are measured by generalized fractal dimensions and multifractal spectra, which are used to achieve the optimal MSP detection. Although, the proposed algorithm is robust to normal and pathological images in comparison with the other existing techniques, the proposed method found MSP as straight, but not exactly planar. It might be a significant MSP detection method, if the non-planar surface that separates the two hemispheres. A possible future work is to extend the multifractal based method for detecting the curved MSP.

	Zhang et al.	Rupper et al.	Liu et al.	Proposed method

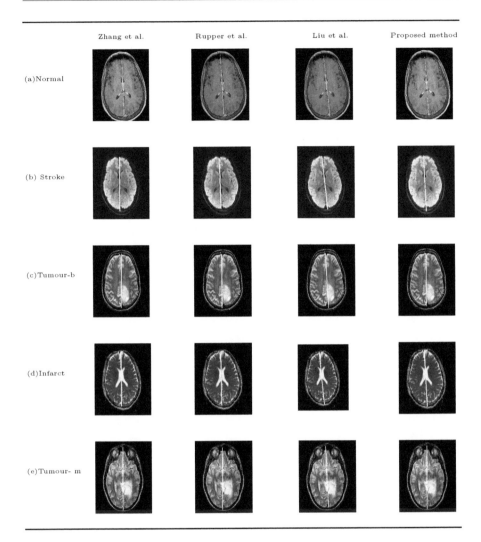

Figure 6.12: Comparison between extracted MSP by the proposed method and three state-of-the-art methods on MRI images: a) Normal b) Stroke c) Tumour-benign d) Infarct e) Tumour-malignant.

6.6.4 Conclusion

The proposed algorithm has included the generalized fractal dimensions as a symmetric measure and the multifractal spectra to refine the optimal mid-sagittal plane. The proposed algorithm has combined GFD and multifractal spectra for a more accurate and efficient extraction of mid-sagittal plane from brain MRI. The proposed algorithm has comprehensively tested on various normal and pathological brain MRI samples with various modalities. Experiments using a large and heterogeneous dataset have shown that the proposed method provided the highest accuracy and precision better than the three state-of-the-art methods, and assessed by the performance evaluation measures. The method presented in this chapter provided the best estimate of the mid-sagittal plane for numerous applications in computer assisted diagnosis.

References

[1] M.V Smoluchowski, Versucheiner Mathematischen Theorie der Koagulations Kinetic Kolloider Lousungen, Z. Phys. Chem. 92, 215, 1917.

[2] S. Chandrasekhar, Stochastic problems in physics and astronomy, Rev. Mod. Phys. 15, 1, 1943.

[3] H.E. Hurst, Long-term storage capacity of reservoirs, Trans. of Am. Soc. of Civil Engg. 116, 1951.

[4] Y.L. Luke, The Special Functions and their Approximations I, New York: Academic Press, 1969.

[5] H.E. Stanley, Introduction to Phase Transitions and Critical Phenomena, Oxford University Press, Oxford, New York, 1971.

[6] S. Karlin and H.M. Taylor, A First Course in Stochastic Processes, 2nd edn. Academic Press, New York, 1975.

[7] W. Rudin, Principles of Mathematical Analysis, 3rd edn., McGraw-Hill Book company, New Delhi, 1976.

[8] B.B. Mandelbrot, Fractals, Fractal Cities, Freeman, San Francisco, 1977.

[9] S.K. Frielander, Smoke, Dust and Haze: Fundamentals of Aerosol Behavior, Wiley, New York, 1977.

[10] P.G. de Gennes, Scaling Concepts in Polymer Physics, Cornell University Press, Ithaca, NY, 1979.

[11] J.E. Hutchinson, Fractals and self similarity, Indiana University Mathematics Journal, 30, 713–747, 1981.

[12] T.M. Apostol, Mathematical Analysis, 2nd edn. Addison-wesley publishing company, London, 1981.

[13] B.B. Mandelbrot, The Fractal Geometry of Nature, W.H. Freeman and Company, New York, 1983.

[14] P. Grassberger and I. Procaccia, Measuring the Strangeness of Strange Attractors, Physica D 9, 189–208, 1983.

[15] P. Grassberger, Generalized dimensions of strange attractors, Physics Letters A, 97, 227–320, 1983.

[16] H.G.E. Hentschel and I. Procaccia, The infinite number of generalized dimensions of fractals and strange attractors, Physica D: Nonlinear Phenomena, 8(3), 435–444, 1983.

[17] T. Viscek and F. Family, Dynamic scaling for aggregation of clusters, Phys. Rev. Lett. 52, 1669, 1984.

[18] R.M. Ziff and E.D. McGrady, The kinetics of cluster fragmentation and depolymerisation, J. Phys. A: Math. Gen. 18, 3027–3037, 1984.

[19] R.M. Ziff and E.D. McGrady, The kinetics of cluster fragmentation and depolymerisation, J. Phys. A: Math. Gen., 18(5), 3027, 1985.

[20] P.G.J. van Dongen and M.H. Ernst, Dynamic scaling in the kinetics of clustering, Phys. Rev. Lett. 54, 1396, 1985.

[21] R.M. Ziff and E.D. McGrady, Kinetics of polymer degradation, Macromolecules, 19(10), 2513–2519, 1986.

[22] C. Amitrano, A. Coniglio and F. di Liberto, Growth probability distribution in kinetic aggregation processes, Phys. Rev. Lett. 57, 1016, 1986.

[23] J. Feder, Fractals, New York: Plenum, New York, 1988.

[24] R.M. Bethea and R.R. Rhinehart, Applied Engineering Statistics, Marcel Dekker, Inc., New York, NY, 1991.

[25] C.-K. Peng, S.V. Buldyrev, A.L. Goldberger, S. Havlin, F. Sciortino, M. Simons and H.E. Stanley, Long-range correlations in nucleotide sequences, Nature 356, 168–171, 1992.

[26] P.L. Krapivsky, Kinetics of random sequential parking on a line, J. Stat. Phys. 69, 135–150, 1992.

[27] M.F. Barnsley, Fractals Everywhere, 2nd edn. Academic Press, USA, 1993.

[28] P.R. Massopust, Fractal Functions, in Fractal Surfaces and Wavelets, Academic Press, San Diego, 1994.

[29] P.L. Krapivsky and E. Ben-Naim, Multiscaling in stochastic fractals, Phys. Lett. A 196, 168–172, 1994.

[30] A. Bunde and S. Havlin, Fractals in Science, Springer, Berlin, 1994.

[31] P.L. Krapivsky and E. Ben-Naim, Multiscaling in stochastic fractals, Phys. Lett. A, 196, 168, 1994.

[32] P.L. Krapivsk and E. Ben-Naim, Scaling and multiscaling in models of fragmentation, Phys. Rev. E, 50, 3502, 1994.

[33] G.J. Rodgers and M.K. Hassan, Fragmentation of particles with more than one degree of freedom, Phys. Rev. E 50, 3458–3463 1994.

[34] A. Renyi, On a new axiomatic theory of probability, Acta Mathematica Hungarica, 6, 285–335, 1995.

[35] M.K. Hassan and G.J. Rodgers, Models of fragmentation and stochastic fractals, Phys. Lett. A, 208, 95–98, 1995.

[36] M. Rao, S. Sengupta and H.K. Sahu, Kinematic scaling and crossover to scale invariance in martensite growth, Phys. Rev. Lett. 75, 2164, 1995.

[37] G.I. Barenblatt, Scaling, Self-similarity, and Intermediate Asymptotics, Cmpridge University Press, 1996.

[38] H.E. Stanley in *Fractals and Disordered Systems* eds. Bunde A and Havlin S, New York: Springer, 1996.

[39] P.L. Krapivsky and E. Ben-Naim, Kinematic scaling and crossover to scale invariance in martensite growth, Phys. Rev. Lett., 76, 3234, 1996.

[40] M.K. Hassan and G.J. Rodgers, Multifractality and multiscaling in two dimensional fragmentation, Phys. Lett. A, 218, 207–211, 1996.

[41] M.K. Hassan, Multifractality and the shattering transition in fragmentation processes, Phys. Rev. E 54, 1126–1133, 1996.

[42] M.K. Hassan, Fractal dimension and degree of order in sequential deposition of a mixture of particles, Physical Review E, 55(5), 5302, 1997.

[43] C.E. Shannon, The Mathematical Theory of Communication, University of Illinois Press, Champaign, IL, 1998.

[44] Lui Lam, Nonlinear Physics for Beginners, World Scientific, Singapore, 1998.

[45] N.A. Salingaros and B.J. West, A universal rule for the distribution of sizes, Environment and Planning B: Planning and Design 26(6), 909–923, 1999.

[46] P. Asvestas, G.K. Matsopoulos and K.S. Nikita, Estimation of fractal dimension of images using a fixed mass approach, Pattern Recognition Letters, 20, 347–354, 1999.

[47] T. Schreiber and A. Schmitz, Surrogate Time Series, Physica D, 142, 346–382, 2000.

[48] E. Ben-Naim and P.L. Krapivsky, Stochastic aggregation: rate equations approach, J. Phys. A: Math. Gen., 33, 547, 2000.

[49] S.H. Strogatz, Nonlinear Dynamics And Chaos: With Applications To Physics, Biology, Chemistry, And Engineering, Persue Book Publishing, 2001.

[50] J.W. Kantelhardt, E. Koscielny-Bunde, H.H.A. Rego and S. Havlin, Detecting long-range correlations with detrended fluctuation analysis, Physica A 295, 441454, 2001.

[51] Y. Liu, R.T. Collins and W.E. Rothfus, Robust midsagittal plane extraction from normal and pathological 3-D neuroradiology images, IEEE Transactions on Medical Imaging, 20(3), 2001.

[52] S. Redner, A Guide to First-Passage Processes, Cambridge University Press, Cambridge, 2001.

[53] N.G. van Kampen, Stochastic Processes in Physics and Chemistry, North-Holland, Amsterdam, 2001.

[54] M.K. Hassan and J. Kurths, Transition from random to ordered fractals in fragmentation of particles in an open system, Phys. Rev. E, 64(1), 016119, 2001.

[55] M.K. Hassan and J. Kurths, Can randomness alone tune the fractal dimension? Physica A, 315, 342–352, 2002.

[56] John C. Russ, The Image Processing Handbook, 4th ed., CRC Press, London, 2002.

[57] K.J. Falconer, Fractal Geometry: Mathematical Foundations and Applications, 2nd. edition, John Wiley & Sons Ltd., England, 2003.

[58] M.E.J. Newman, SIAM Review, 45, 167, 2003.

[59] C. Jingdong, Filtering Techniques for Noise Reduction and Speech Enhancement, Chapter in Adaptive Signal Processing: Applications to Real-World Problems, Springer Berlin Heidelberg, 129–154, 2003.

[60] T. Gautama, D.P. Mandic and M.M.V. Hulle, The delay vector variance method for detecting determinism and nonlinearity in time series, Physica D, 190, 167–176, 2004.

[61] S.N. Majumdar, D.S. Dean and P.L. Krapivsky, Understanding search trees via statistical physics, *Pramana - J. Phys.*, 64(6), 1175–1189, 2005.

[62] Y.S. Liang and W.Y. Su, The relationship between the fractal dimensions of a type of fractal functions and the order of their fractional calculus, Chaos, Solitons and Fractals, 34, 682–692, 2007.

[63] K.R. Castleman, Digital Image Processing, Pearson Education India, 1st ed., 2007.

[64] G. Edgar, Measure, Topology, and Fractal Geometry, 2nd edition, Springer, New York, 2008.

[65] Y. Zhang and Q. Hu, A PCA-based approach to the representation and recognition of MR brain midsagittal plane images, 30th Annual International Conference of the IEEE-EMBS, 3916–3919, 2008.

[66] M.K. Hassan and M.Z. Hassan, Condensation-driven aggregation in one dimension, Phys. Rev. E 77, 061404, 2008.

[67] G.W. Delaney, S. Hutzler and T. Aste, Relation between grain shape and fractal properties in random apollonian packing with grain rotation, Phys. Rev. Lett., 101, 120602, 2008.

[68] M.K. Hassan and M.Z. Hassan, Emergence of fractal behavior in condensation-driven aggregation, Phys. Rev. E, 79(2), 021406, 2009.

[69] P.A. Varotsos, N.V. Sarlis and E.S. Skordas, Detrended fluctuation analysis of the magnetic and electric field variations that precede rupture, Chaos 19, 023114, 2009.

[70] M.K. Hassan and M.Z. Hassan, Emergence of fractal behavior in condensation-driven aggregation, Phys. Rev. E, 79, 021406, 2009.

[71] J. Aguirre, R.L. Viana and M.A.F. Sanjuan, Rev. Mod. Phys. 81, 333, 2009.

[72] Jian Li, Qian Du and Caixin Sun, An improved box-counting method for image fractal dimension estimation, Pattern Recognition, 42, 2460–2469, 2009.

[73] R.K.P. Zia, E.F. Redish and S.R. McKay, Am. J. Phys., 77, 614, 2009.

[74] P.R. Massopust, Interpolation and Approximation with Splines and Fractals, Oxford University Press, New York, 2010.

[75] M.K. Hassan, M.Z. Hassan and N.I. Pavel, Scale-free network topology and multifractality in a weighted planar stochastic lattice, New J. Phys., 12, 093045, 2010.

[76] MRI Image Database, website: http://www.cma.mgh.harvard.edu/ibsr./

[77] MRI Image Database, website: http://www.bic.mni.mcgill.ca/brainweb./

[78] S.X. Liu, J. Kender, C. Mielinska and A. Laine, Employing symmetry features for automatic misalignment correction in neuroimages, Journal of Neuroimaging, 21, 15–33, 2011.

[79] G.C.S. Ruppert, L.A. Teverovskiy, C.P. Yu, A.X. Falcao and Y. Liu, A new symmetry-based method for mid-sagittal plane extraction in neuroimages, IEEE International Symposium on Biomedical Imaging: From Nano to Macro, 285–288, 2011.

[80] M.K. Hassan, M.Z. Hassan and N.I. Pavel, J. Phys: Conf. Ser, 297, 012010, 2011.

[81] A. Biswas, T.B. Zeleke and B.C. Si, Multifractal detrended fluctuation analysis in examining scaling properties of the spatial patterns of soil water storage, Nonlin. Processes Geophys., 19, 227238, 2012.

[82] S.A. Jayasuriya, A.W.C. Liew and N.F. Law, Brain symmetry plane detection based on fractal analysis, Computerized Medical Imaging and Graphics, 37, 568–580, 2013.

[83] M.K. Hassan, M.Z. Hassan and N. Islam, Emergence of fractals in aggregation with stochastic self-replication, Phys. Rev. E, 88(4), 042137, 2013.

[84] R. Uthayakumar and A. Gowrisankar, Generalized Fractal Dimensions in Image Thresholding Technique, Information Sciences Letters, Natural Sciences, 3(3), 125–134, 2014.

[85] M.K. Hassan, N.I. Pavel, R.K. Pandit and J. Kurths, Chaos, Solitons & Fractals, 60, 31–39, 2014.

[86] R. Uthayakumar and A. Gowrisankar, Generation of Fractals via Self-Similar Group of Kannan Iterated Function System, Applied Mathematics & Information Sciences, Natural Sciences, 9(6), 3245–3250, 2015.

[87] R. Uthayakumar and A. Gowrisankar, Attractor and self-similar group of generalized fuzzy contraction mapping in fuzzy metric space, Cogent Mathematics, Taylor & Francis 2(1): 1024579, 1–12, 2015.

[88] A. Gowrisankar and R. Uthayakumar, Fractional calculus on fractal interpolation for a sequence of data with countable iterated function system, Mediterranean Journal of Mathematics, Springer, 13(3), 3887–3906, 2016.

[89] R. Uthayakumar and A. Gowrisankar, Mid-sagittal plane detection in magnetic resonance image based on multifractal techniques, IET Image Processing, 10(10), 751–762, 2016.

[90] A. Gowrisankar, Generation of fractal through iterated function systems, Ph.D. Thesis, The Gandhigram Rural Institute (Deemed to be University), 2016.

[91] Y.S. Liang and Q. Zhang, A type of fractal interpolation functions and their fractional calculus, Fractals, 24(2), 1650026, 2016.

[92] F.R. Dayeen and M.K. Hassan, Chaos, Solutions & Fractals 91 228, 2016.

[93] R.C. Gonzalez, R.E. Woods and S.L. Eddins, Digital Image Processing Using MATLAB, 2nd ed., McGraw Hill Education, 2017.

[94] A. Gowrisankar and M. Guru Prem Prasad, Riemann-Liouville calculus on quadratic fractal interpolation function with variable scaling factors, The Journal of Analysis, Springer, 1–7, 2018.

[95] A. Gowrisankar and D. Easwaramoorthy, Local countable iterated function systems, ICAMS-2017—Trends in Mathematics—Springer Book Series, 1, 169–175, 2018.

[96] C. Raja Mohan, A. Gowrisankar, R. Uthayakumar and K. Jayakumar, Morphology dependent electrical property of chitosan film and modeling by fractal theory, The European Physical Journal Special Topics, 228, 233–243, 2019.

[97] M.K. Hassan, Is there always a conservation law behind the emergence of fractal and multifractal, Eur. Phys. Journal ST, 228(1), 209–232, 2019.

Index

T - #0854 - 101024 - C206 - 234/156/9 - PB - 9781032083513 - Gloss Lamination